# FUNDAMENTALS OF
# HOT WIRE ANEMOMETRY

This book provides a clear and comprehensive summary of the theory and practice of the hot wire anemometer, an instrument used to measure the speed of fluid flow. Many techniques and uses of this instrument, which until now have only appeared in technical journals, are discussed in detail. The author considers such topics as probe fouling, probe design, and circuit design, as well as the thermodynamics of heated wires and thin films. He also discusses measurements of turbulence, shear flows, vorticity, temperature, combined temperature and velocity, two-phase flows, and compressible flows, as well as measurements in air, water, mercury, blood, glycerine, oil, luminous gases, and polymer solutions. The book concludes with a section on the pulsed wire anemometer and other wake-sensing anemometers.

*Fundamentals of Hot Wire Anemometry* is written at the advanced undergraduate level and assumes a familiarity with basic fluid mechanics. However, mathematical descriptions occur near the end of each chapter, thus allowing those with a limited mathematical background access to the practical details at the beginning of each chapter. The volume will be useful to students, teachers, and researchers in fluid mechanics, and will serve as a handy reference to all users of the hot wire anemometer.

# FUNDAMENTALS OF HOT WIRE ANEMOMETRY

CHARLES G. LOMAS
Rochester Institute of Technology

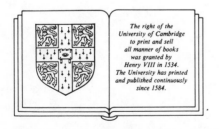

The right of the
University of Cambridge
to print and sell
all manner of books
was granted by
Henry VIII in 1534.
The University has printed
and published continuously
since 1584.

CAMBRIDGE UNIVERSITY PRESS

*Cambridge*
*London New York New Rochelle*
*Melbourne Sydney*

To my wife, Arletta,
and our daughter, Kathy Mullen

Published by the Press Syndicate of the University of Cambridge
The Pitt Building, Trumpington Street, Cambridge CB2 1RP
32 East 57th Street, New York, NY 10022, USA
10 Stamford Road, Oakleigh, Melbourne 3166, Australia

First published 1986

Printed in the United States of America

*Library of Congress Cataloging in Publication Data*

Lomas, Charles G. (Charles Gardner), 1934–

Fundamentals of hot wire anemometry.

Bibliography: p.
1. Fluid dynamic measurements.   2. Hot-wire
anemometer.   I. Title.
TA357.L595   1985   532'.0532   85-12739

*British Library Cataloguing in Publication Data*

Lomas, Charles G.

Fundamentals of hot wire anemometry.

1. Hot wire anemometer
I. Title
620.1'064   TA357
ISBN 0 521 30340 0

# CONTENTS

v

# PREFACE

Over a number of years I observed that those who used the hot wire anemometer for velocity measurements in air were not aware of its many other
applications. In addition, their knowledge was often limited to that obtained
from instruction manuals and co-workers. This prompted me to write a book
that would contain theoretical and practical information that otherwise could
be acquired only by spending many hours researching the technical
literature.

Readers will need an undergraduate engineering background, but the information will be accessible both to a research engineer designing a test
program and to the technician who may do the testing.

The reader will find here the theoretical background for the measurements
of velocity, turbulence, compressible flows, vorticity, temperature, concentration, and two-phase flows, as well as practical information about probe
design and measurements in fluids, such as air, water, polymer solutions,
mercury, blood, glycerine, oil, and luminous gases. To appeal to the enthu-
siast, novelties are included, such as a hot wire probe that costs almost
nothing and is constructed in less than 1 minute by using one tool – a pair
of pliers – and a low-velocity calibration technique in air using soap bubbles.
Illustrations of handmade probes have been chosen to show the excellent
results possible with simple tools and a little patience.

The manufacturers of hot wire anemometers were very helpful in contrib-
uting illustrations of commercially available probes. The following manufacturers graciously donated both time and information: Deltalab (France),
Dantec Elektronik, formerly DISA (Denmark), Malvern Instruments (England), Prosser Scientific Instruments, Ltd. (England), and TSI, Inc. (United
States).

This book could not have been written without the help of many individ-
uals. Thanks go to Robert Bechov, Charlie Best, Ray Bowles, K. Bremhorst,
K. Brodowicz, G. L. Brown, Preben Buchhave, H. H. Bruun, Frank Champagne, Orla Christensen, G. Comte-Bellot, Stanley Corrsin, P. O. A. L.
Davies, J. W. Delleur, C. Forbes Dewey, G. E. Dix, H. Eckelmann, B. S.
Fedders, R. Fekkes, H. Fiedler, Kenneth Forman, John Foss, Brent Gallagher, Benjamin Gebhart, F. B. Gessner, Preben Gundersen, E. G. Hauptmann, Edward Hayes, Jim Hill, Ron Humphrey, Finn Jorgensen, W.

Katscher, N. W. M. Ko, William Kolbeck, Michael Kotas, Leslie Kovasznay, Ed Krick, Richard Kronauer, A. J. Laderman, Soren Larsen, Paul Libby, H. W. Liepmann, S. C. Ling, Paul Lykoudis, J. McQuaid, Raul McQuivey, Colin Marks, Edward Merrill, Patrice Mestayer, Jerry Miller, G. Mollenkopf, Phil Mulhearn, Robert Nerem, W. Neuerburg, B. G. Newman, Simeon Oka, A. E. Perry, J. Pichon, Felix Pierce, A. F. Polyakov, J. N. Prosser, Dale Pullen, C. G. Rasmussen, Claude Rey, F. P. Ricou, Frederick Roos, Anatol Roshko, Dirse Sallet, Virgil Sandborn, R. "Sat" Sathyakumar, Gordon Schacher, Tony Seed, L. Shemer, Robert Stewart, Bob Suhoke, D. Trudgill, W. Willmarth, John Wills, and Scott Zosel. I also wish to thank my wife, Arletta; Anne Blake, who assisted with editing; and Betty Bullock, who typed much of the preliminary draft. Mary DeGuzman and Leah Herman provided moral support. Special thanks go to Russell Hahn, Rhona Johnson, Peter-John Leone, and David Tranah of Cambridge University Press for their help in bringing this manuscript to completion as well as to Tom Whipple, who edited the final manuscript.

Charles G. Lomas
*Rochester, New York*

# NOTATION

**Symbols**

| | |
|---|---|
| $a$ | overheat ratio |
| $A$ | area |
| $A$ | Wheatstone bridge ratio |
| $b$ | yaw parameter |
| $B$ | magnetic flux density |
| $c$ | specific heat |
| $C$ | fouling factor |
| $d$ | diameter |
| $e$ | fluctuating component of voltage |
| $e_o$ | Wheatstone bridge off-balance voltage |
| $E$ | voltage |
| $f$ | frequency |
| fr | frequency response |
| $g$ | acceleration of gravity |
| $g$ | feedback amplifier transconductance |
| Gr | Grashof number |
| $h$ | coefficient of convective heat transfer |
| $h$ | distance |
| $h$ | pitch factor |
| Ha | Hartmann number |
| $i$ | fluctuating component of current |
| $I$ | electrical current |
| $k$ | thermal conductivity |
| $k$ | yaw factor |
| $K$ | constant |
| $K$ | slope of the linearized calibration curve |
| Kn | Knudsen number |
| $l$ | length |
| $L$ | fixed length |
| $L$ | inductance |
| $M$ | Mach number |
| $n$ | exponent used in King's law |
| $n$ | molecular density |
| $N$ | magnetic interaction parameter |

ix

| | |
|---|---|
| Nu | Nusselt number |
| $p$ | distance |
| $P$ | pressure |
| Pe | Peclet number |
| Pr | Prandtl number |
| $q$ | heat transfer rate |
| $r$ | fluctuating component of electrical resistance |
| $r$ | radius |
| $R$ | electrical resistance |
| Ra | Rayleigh number |
| Re | Reynolds number |
| $S$ | sensitivity |
| $S$ | shear factor |
| Sn | Strouhal number |
| $t$ | fluctuating component of temperature |
| $t$ | time |
| $T$ | temperature |
| $T$ | time of flight |
| $u$ | fluctuating component of velocity |
| $U$ | X-component of velocity |
| $v_*$ | friction velocity |
| $x$ | characteristic length |
| $x$ | horizontal distance |
| $X$ | Sajben $X$-factor |
| $y$ | vertical distance |
| $z$ | lateral distance |
| $\alpha$ | temperature coefficient of resistivity |
| $\alpha$ | thermal diffusivity |
| $\alpha$ | angle of inclination of the velocity vector |
| $\beta$ | volume coefficient of expansion |
| $\beta$ | angle of inclination of the velocity vector |
| $\gamma$ | angle of inclination of the velocity vector |
| $\gamma$ | specific heat ratio |
| $\gamma$ | fluctuating component of molal concentration |
| $\Delta$ | small change in value |
| $\epsilon$ | emissivity |
| $\epsilon$ | relative error |
| $\eta$ | dimensionless length |
| $\theta$ | fluctuating component of temperature |
| $\theta$ | temperature difference |
| $\theta$ | yaw angle |
| $\lambda$ | mean free path length |
| $\Gamma$ | circulation |
| $\Gamma$ | concentration |
| $\mu$ | absolute viscosity |

| | |
|---|---|
| $\nu$ | kinematic viscosity |
| $\rho$ | density |
| $\rho_r$ | resistivity |
| $\sigma$ | Stefan–Boltzmann constant |
| $\sigma$ | electrical conductivity |
| $\tau$ | time constant |
| $\tau$ | shear stress |
| $\phi$ | phase angle |
| $\phi$ | pitch angle |
| $\psi$ | normalized voltage |
| $\psi$ | roll angle |
| $\omega$ | circular frequency |
| $\omega$ | rolloff frequency |
| $\omega$ | vorticity |

**Subscripts**

| | |
|---|---|
| adj | adjustable |
| $a$ | apparent |
| $a$ | air |
| $b$ | bridge |
| $c$ | cable |
| $c$ | corrected |
| $c$ | critical |
| cca | constant current anemometer |
| cta | constant temperature anemometer |
| cl | centerline |
| conc | concentration |
| $e$ | equilibrium |
| eff | effective |
| eq | equivalent |
| $f$ | fluid |
| $g$ | gas |
| $i$ | impurities |
| $m$ | mean |
| $m$ | measured |
| $m$ | mixture |
| max | maximum |
| min | minimum |
| $n$ | arbitrary orthogonal coordinate |
| $o$ | reference or stagnation conditions |
| off | offset |
| oven | oven |
| $p$ | probe |
| $P$ | constant pressure |
| $q$ | quartz |

| | |
|---|---|
| $s$ | sensor |
| $s$ | solvent |
| $s$ | streamwise |
| sub | substrate |
| sur | surroundings |
| $t$ | arbitrary orthogonal coordinate |
| $T$ | temperature |
| temp | temperature |
| vel | velocity |
| $w$ | wall |
| 1 | conditions upstream of a normal shock wave |
| 2 | conditions downstream from a normal shock wave |
| $\infty$ | free stream conditions |

# 1 INTRODUCTION

Hot wire anemometers have been used since the late 1800s, when experimentalists in fluid mechanics built their own rudimentary constant current anemometers. Because no commercial equipment was available, all improvements were made by the scientists themselves. Eventually, electronic amplification and shaping networks were added, and the constant current anemometer became a sophisticated, high-frequency-response research instrument. The appearance of commercial constant current anemometers, and later of commercial constant temperature anemometers, coincided with a major growth in popularity of these instruments. Today the hot wire anemometer is used in research laboratories throughout the world.

Besides the use of the hot wire anemometer in air, measurements can be made in other fluids, such as fresh water, salt water, polymer solutions, blood, mercury, glycerine, oil, freon, and luminous gases. In addition, the hot wire anemometer can be used to determine the direction and speed of a fluid, to make turbulence measurements, to make measurements in compressible flows, and to measure fluid temperature. Special techniques allow the hot wire anemometer to measure gas mixture concentrations and to make two-phase flow measurements as well.

The hot wire anemometer also has excellent frequency response; an upper frequency limit of 400 kHz is common for commercially available instruments. Only the laser Doppler velocimeter, which at this writing is four to five times more expensive for an equivalent system, can compete in this respect. In addition, the hot wire anemometer has excellent sensitivity at low velocity, good spatial resolution, and an output signal in the form of a voltage difference for convenient data analysis. All in all, the hot wire anemometer is one of the most flexible instruments available for research in fluid mechanics.

The name *hot wire anemometer* implies using a heated wire to make velocity measurements in air only. The word *anemometer* is inaccurate because the instrument is used in a variety of fluids, and the term *hot wire* is misleading because probes using a heated metal film are popular. This nomenclature was adopted in the early 1900s, when hot wire probes were used in air measurements only. Times have changed, but the name remains.

Every hot wire anemometer, regardless of type, contains the same basic parts: a probe with its cable, and an electronics package. A typical hot wire probe is illustrated in Figure 1.1.

1

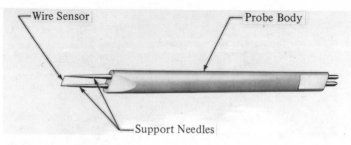

Figure 1.1. A typical hot wire probe. This is the most common and least expensive type; it has a single-wire sensor attached to the tips of two support needles and a connector at the other end to allow quick detachment for cleaning, recalibration, or replacement. Reprinted with permission from Dantec Elektronik.

The sensor of the typical hot wire probe is a wire, usually made of tungsten or platinum, about 1 mm long and 5 μm in diameter. It may have a thin plating of a different metal. The diameter of the wire is much less than that of a human hair, which has a diameter of about 85 μm. This means that the sensor cannot be easily seen with the unaided eye of the inexperienced user. The sensor is attached between the tips of two support needles by arc welding or soldering, and is electrically heated. It is convection cooled by the fluid passing over it, and this cooling effect is a measure of the fluid velocity. The probe body is usually made of epoxy or ceramic material or fabricated from a metal tube potted with epoxy. An electrical connector is often located at the other end of the probe body to allow easy removal and replacement of the probe. The contacts of the connector are sometimes plated with gold to reduce resistance, and the connector is usually designed to be watertight.

The sensor of a hot film probe is usually made of nickel or platinum deposited in a thin layer onto a backing material, such as quartz, and connected to the electronics package by leads attached to the ends of the film. A thin protective coating of quartz or other material is usually deposited over the film to prevent damage by abrasion or chemical reaction. The wedge hot film probe illustrated in Figure 1.2 is quite popular. This probe is made from a quartz rod ground to a wedge at one end, and the metal film is deposited in a strip along the knife edge of the wedge.

Three types of electronics packages are used, each controlling the sensor heating current in a different way. The most common is the constant temperature anemometer, which supplies a sensor heating current that varies with the fluid velocity to maintain constant sensor resistance and, thus, constant sensor temperature. Less often used is the constant current anemometer, which supplies a constant heating current to the sensor. A third type, the pulsed wire anemometer, measures velocity by momentarily heating a wire to heat the fluid around it. This spot of heated fluid is convected downstream to a second wire that acts as a temperature sensor. The time of flight of the hot spot is inversely proportional to the fluid velocity.

Quartz Rod

Hot Film Sensor

Figure 1.2. A wedge hot film probe. This probe is the hot film analog of the standard single-sensor hot wire probe shown in Figure 1.1. A quartz rod having a "chisel" shape is plated with a thin film of metal on the leading edge. Reprinted with permission from TSI, Inc.

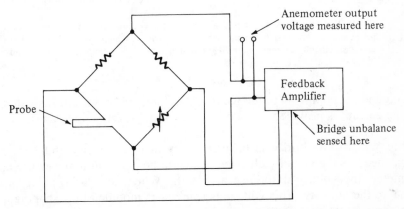

Anemometer output
voltage measured here

Feedback
Amplifier

Probe

Bridge unbalance
sensed here

Figure 1.3. The block diagram of a constant temperature anemometer. A hot wire probe acts as one resistor in the Wheatstone bridge, and the feedback amplifier automatically adjusts the current to maintain bridge balance.

The electronics package of the constant temperature anemometer contains a Wheatstone bridge circuit with the sensor as one arm of the bridge, as shown in Figure 1.3; two fixed resistors and one adjustable resistor complete the circuit.

A differential feedback amplifier senses the bridge unbalance and adds current to hold the sensor temperature constant. Before the system is placed in operation, the adjustable resistor is set to a value larger than would be required to balance the bridge. When power is applied, the feedback amplifier increases the sensor heating current, causing the sensor temperature to rise and increase the sensor resistance until the bridge becomes balanced. An increase in velocity cools the sensor and unbalances the bridge. This causes the feedback amplifier to increase the sensor heating current and to bring the bridge back into balance. Since the feedback amplifier responds rapidly, the sensor temperature remains virtually constant as the velocity changes. The voltage difference across the bridge is proportional to the fluid velocity.

The constant current anemometer may contain both a Wheatstone bridge

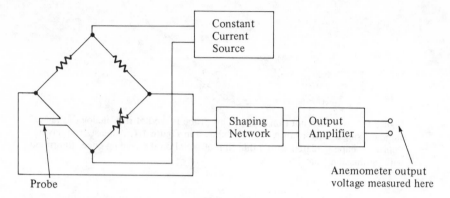

Figure 1.4. The block diagram of a constant current anemometer. A hot wire probe acts as one resistor in the Wheatstone bridge, which is powered by a constant current source. The bridge unbalance voltage is shaped and amplified before reaching the recording device.

circuit and an amplifer, but the feedback technique is not used. Instead, the bridge current is provided by a constant current power supply, as shown in Figure 1.4.

The Wheatstone bridge is balanced only at one velocity and becomes unbalanced as the velocity changes. As with the constant temperature anemometer, the voltage difference across the Wheatstone bridge is proportional to the velocity. This voltage is sometimes modified by a shaping network and amplifier to improve the frequency response.

We now give some definitions that are basic to hot wire anemometry. The first, the sensitivity of the hot wire anemometer to changes in fluid speed, temperature, or direction of the mean velocity vector, is defined as a derivative of the anemometer output voltage with respect to the fluid property under consideration. For example, the velocity and temperature sensitivities, $S_{vel}$ and $S_{temp}$, can be expressed as

$$S_{vel} = \frac{\partial E}{\partial U}$$

and

$$S_{temp} = \frac{\partial E}{\partial T}$$

where $E$ is the anemometer output voltage, $U$ is the velocity, and $T$ is the fluid temperature. Because the output voltage is a function of both fluid velocity and temperature, the chain rule gives

$$dE = \frac{\partial E}{\partial U}\, dU + \frac{\partial E}{\partial T}\, dT \tag{1.1}$$

where the sensitivities appear as factors in this equation.

The concept of sensor operating temperature is important because it influences both the life of the probe and its sensitivity to velocity and ambient temperature changes. The sensor temperature is usually expressed as a ratio, called the *overheat ratio, a*, and defined as either a resistance ratio

$$a_1 = \frac{R_s}{R_f}$$

or a resistance difference ratio

$$a_2 = \frac{R_s - R_f}{R_f}$$

where $R_s$ is the resistance of the heated sensor at its operating temperature, and $R_f$ is the resistance of the sensor at the temperature of the ambient fluid. The first definition is used in this book.

A different definition for sensor operating temperature is sometimes used by those who make blood flow measurements. Because blood can only be heated a few degrees centigrade above body temperature without damage, the overheat concept is sometimes expressed as an "overheat temperature" of, for example, 5°C. This means the adjustable resistor in the Wheatstone bridge is set to a value that allows the sensor temperature to be held at 5°C above ambient blood temperature.

# 2 USING THE HEATED SENSOR PROBE

In this chapter we discuss probe life and how to extend it, probe fouling, probe support, and considerations such as vortex shedding from cylindrical sensors, directional sensitivity, and the interference that may exist between closely spaced probes. We conclude by discussing design criteria for probes and techniques for making hot wire and hot film probes by hand.

## 2.1 Probe life

Although the usable life of a probe depends upon so many variables that no estimate of it can be given, probe life, and your ability to extend it, is important. For long life the aging process and the reasons for premature failure must be understood.

### Sensor aging

Although one cannot know precisely when a probe will fail, awareness of the aging symptoms allows one to know when a probe is near the end of its life. The most important indicators of sensor aging are total hours of operation, sensor operating temperature used, cold resistance drift, and loss in sensitivity as expressed as a change in the slope of the calibration curve. By keeping track of these indicators, even to the extent of recording each in a log book, one can determine when a probe is near the end of its life.

When a hot wire probe is new, it is virtually unused; only a quick operational check and measurement of its cold resistance have usually been made. When heated, the wire temperature is greatest near the center because heat is lost at the ends by conduction to the support needles. In addition, the electrical resistance of the sensor is greatest at points along the wire where the diameter is least, and this causes hot spots when the sensor is heated. Hot spots will also occur under dirt particles or other fouling material, because the wire will not be cooled well by convection at these points. Finally, oxidation of the wire in air is greatest at hot spots, adding to the probability of burnout or breakage there.

When new, the hot film probe is virtually unused also. When operating, the hot film sensor is hottest at points where less metal is deposited. A particle of dirt attached to the sensor can also cause a hot spot. In addition, the insulation coating may have pinholes, allowing oxidation or chemical

6

etching of the metal film. But even if no pinholes are present, cracks can form in the insulation coating due to expansion and contraction of the metal film underneath. In water a high voltage difference between the sensor and the water may cause a breakdown in the insulation coating and result in electrolytic etching of the sensor and eventual burnout at that spot. An interesting photograph of an aged cylindrical hot film sensor can be found in Delleur, Toebes, and Liu (1968).

### Slowing the sensor aging process

Sensor aging can be noticeably reduced by reducing both the length of time the sensor is heated and the sensor temperature, often without compromising the test results.

In a lengthy test it is not unknown for probes to be left powered for long periods of time when no actual testing is taking place. For example, the interruption of a test due to a malfunction of some other piece of equipment could divert attention away from the probes, which might remain heated for hours. Allowing the probes to remain heated during the lunch hour or over a weekend is not unknown. The obvious solution is to switch off the probe heating current when no measurements are being taken; a front panel "standby" switch is usually provided for this purpose.

A sensor will last longer if it is operated at a sensor temperature as low as possible. Of course, the sensor temperature should not be so low that velocity sensitivity is seriously reduced. All operational checks of the equipment can be conducted at a lower overheat ratio as well.

### Sensor preaging

After discussing ways to slow the aging process, it may come as a surprise to learn that the sensor may be purposely aged before a test commences. This is only done if hostile environmental conditions could cause a change in sensor cold resistance while measurements are taking place. An example is the rapid and continuous cold resistance drift of an unaged sensor in supersonic flow that would invalidate the calibration curve and make accurate measurements impossible.

A probe can be preaged by operating it at the highest overheat ratio and velocity expected during the test. The cold resistance should be checked periodically and preaging terminated when excessive drift has not occurred for several hours.

Film probes without an insulation coating, although seldom used anymore, suffer an initial cold resistance drift when used in air, due to oxidation of the film (Christensen, 1970). After preaging the oxidized surface layer acts as a protective coating to reduce the rate of oxidation and prevent further cold resistance drift.

### Sensor burnout

The skilled person seldom burns out a sensor, although one or two sensors are usually sacrificed in developing these skills. A burnout can be diagnosed, however, and distinguished from other probe malfunctions.

It is easier to diagnose sensor burnout if no hydrodynamic forces are present, as, for example, when burnout occurs during an operational check. The sensor will look intact to the unaided eye, but inspection with a magnifying glass or microscope will reveal a small section missing from along the sensor length. If hydrodynamic forces are present, the sensor may be bent in the flow direction, making the burnout indistinguishable from breakage due to impact by a small particle.

Sensors can be burned out, one after the other, in an attempt to give each an operational check with the overheat ratio set too high. In this case, checking the first probe causes it to burn out, and if the probe is assumed to have been defective beforehand and the other probes tested, each will be burned out in turn – a costly mistake.

Even the experienced person can burn out a sensor if the heating current increases suddenly. This can happen if a probe is connected to an operating electronics package. With no probe attached, the feedback amplifier will send very little current to the bridge because an infinite sensor resistance is detected. The rapid change from infinite to finite sensor resistance as the probe is attached will cause the current to increase suddenly and overshoot, creating a high-amplitude current spike that could damage the sensor. The result can be the same for a momentary open circuit in the probe cable during operation. This type of failure is common with hot wire anemometers of early manufacture and those that are handmade. The probe of a modern hot wire anemometer is protected by a current limiter that prevents the sensor current from rising above an amplitude that could cause burnout. Usually the cutoff amplitude can be set by the user.

Burnout can also occur if one section of the sensor is cooled more than the rest. This can happen if a hot wire probe is used to measure air currents between water waves. In this case a high overheat ratio might be required for adequate sensitivity for the low velocities encountered and to boil off any water that may cling to the sensor after the inevitable immersion. But if a wave strikes the sensor in such a way that one part is immersed while the other part remains in air, the feedback amplifier adds current to replace the heat lost to the underwater portion of the sensor, which causes overheating and burnout of the portion in air.

The sensor of a constant current anemometer is subject to a unique type of burnout. To make velocity measurements in high-speed flow with a constant current anemometer, the probe is placed in the flow field and the sensor heating current set to give adequate sensitivity. But if the fluid speed is inadvertently reduced – for example, when a wind tunnel is turned off at the conclusion of a test – convection will not adequately cool the sensor, and it can burn out.

### Probe breakage
Although even the experienced person damages probes occasionally, the cause can usually be found and the problem cured.

A common cause of breakage by an unskilled person is contact with the fingers. A novice may not realize that a thin wire is located between the tips of the support needles of a hot wire probe and may inadvertently touch these parts, breaking the sensor. If the fingers are wiped along the support needles, the sensor may be completely swept away.

Even more catastrophic than loss of a sensor by improper handling is damage to the sensor and support needles when inserted into holes or placed in a protective container. Although it may not seem possible to hold a probe steady enough to insert it in a small hole without bumping a support needle, this can easily be done with a technique used by those who repair watches: Hold the probe body between the thumb and forefinger and place the remaining fingers and heel of the hand against the wall near the hole. While pressing the hand firmly against the wall, move the probe forward by flexing only the thumb and forefinger. The fingers will shake very little, and the probe is easily inserted. A probe can be inserted in its protective cover by holding the cover in one hand and the probe in the other and pressing the unused fingers and heels of each hand together to reduce shaking as the probe and its cover are moved toward each other. Although not clearly shown in the photograph, the hands are held this way in Figure 2.1 (p. 16), where a hot film probe is being carefully cleaned with a brush.

Another common cause of probe breakage is impact by particles carried along with the fluid. Not only can ordinary room air dust damage the wire sensor, but dust and dirt in wind tunnels and water channels can be a serious problem. Sometimes the simplest procedure is to clean the fluid to see if the breakage rate is reduced. However, in some industrial environments it may not be possible to either clean the fluid or install filters, and a more robust sensor must be used.

Probes mounted on airplanes can be protected from damage caused by airborne dirt particles by installing a shield designed to rotate up in front of the probe (Jacobsen, 1977). The probe is shielded during takeoff and low altitude flight, when dirt particles could damage the sensor. At altitudes where the air is clean, the shield is rotated down to allow measurements to be taken.

Even if the fluid is not contaminated by particles, the hydrodynamic force of the fluid itself can cause the sensor to break, and this possibility is more likely if a large coating of contaminants is present on the sensor, giving it a larger mass and diameter with no increase in strength. A sensor that does not break may still be unusable if the hydrodynamic force causes the wire to stretch and change resistance.

The sensor is more likely to fail if it is not securely soldered or welded to the support needles. Also, if the sensor is overheated when it is soldered or welded in place, the ends could be aged more than the rest of the sensor, causing eventual burnout.

Vibration caused by vortices shed from the sensor or vibration transmitted through the probe support can cause such large wire accelerations that the sensor breaks. Vibration transmitted through the probe body to the support needles can cause them to move relative to one another, resulting in cyclic stretching and eventual breakage of the wire. Kovasznay (1953) found that applying a small drop of rubber cement at the points where the sensor is attached to the support needles acts as a shock absorber to reduce the tendency of the sensor to vibrate.

Optical magnification is needed to diagnose probe breakage. A simple pocket magnifying glass is adequate, but a binocular microscope allows three-dimensional viewing. If you make and repair your own probes, a binocular microscope is probably already available.

No special jigs are needed to hold the probe while viewing it under a microscope. Hold the probe between the fingers and press the hand firmly against the microscope stage. The probe can be held steady and still be rotated and angled for inspection.

One of the easiest problems to diagnose is breakage caused by contact of the support needles with a hard object, if major damage to the support needles occurs. The direction of the bend in the support needles indicates the direction of the impact.

A foolproof diagnosis can be made if an inspection shows the wire to be completely missing and the support needles unharmed. This is characteristic of damage caused by wiping the support needles with the fingers. In no other way could the sensor be completely removed while leaving the support needles unharmed.

More difficult to diagnose is a sensor attached at only one end. This usually indicates the soldered or welded joint was poorly made. If the sensor is broken near a support needle, several conclusions can be drawn. First, excessive heat may have been used to attach the sensor. Or the damage could be caused by vibration or impact by airborne particles. If the sensor is broken near the support needles and bent in a direction other than downstream, it is likely that the sensor was broken and pulled away due to contact with the fingers.

Hot film probes, although more rugged than hot wire probes, are not immune to breakage. Careless wiping of the tip or using it to puncture blood vessels during blood flow measurements can damage the sensor. The insulation coating can also be worn away gradually, and McQuivey (1972) reports that the quartz coating of hot film probes used near the sandy bottom of an estuary were worn away after about 15 hours of exposure.

### Repairing probes

It has been said that one remains a novice in hot wire anemometry until the first probe has been broken, and whether or not this is true, probe breakage is so common that a quick and easy method of repair is desirable.

There are some probes and some types of damage that cannot be repaired. Generally, for a wire probe, severe damage that breaks off one or more support needles close to the probe body is irreparable, because a support needle cannot easily be replaced. Also, one cannot repair the burned-out sensor of most hot film probes or major damage to the substrate.

If a probe is damaged, one must decide whether it should be repaired in the laboratory or returned to the factory for repair. Factory repairs are performed by trained people, but this is offset by repair costs that may be as much as half the cost of a new probe, in addition to a relatively slow turnaround time. A significant part of the cost and time for a factory repair is the handling, packaging, and paperwork needed to return the probe to you. The time required to replace the sensor on a wire probe, for example, may be no longer than the time required to type the invoice and shipping label. Also, the turnaround time for the repair may be dictated more by the speed of the probe through the mail than the time the probe remains at the factory.

For an extensive test program where probe failure could result in expensive delays, you may want to learn to repair your own probes. Hot wire probe repair equipment can be purchased from the manufacturer, or, if funds are lacking, it can be made in the laboratory.

The hot wire probe repair equipment sold commercially is often identical to that used at the factory for repairs. It often consists of a micromanipulator to position a stretched wire across the support needle tips and to position the welding electrode at each tip to weld the joint. Also included may be a spot-welding current generator. An attachment point for a binocular microscope is usually provided to allow viewing the repair while doing it.

A wire sensor is repaired in one of these commercially available devices in the following way: First insert the probe into its holder on the micromanipulator. Next, place the small spool of sensor wire on its fixture, unroll a length of the wire, and clamp it between the two holders. The most difficult part of this procedure may be finding the end of the fine wire on the spool! Then, while looking through the microscope, position the wire holders to place the wire across the tips of both support needles. Sandwich the wire between the electrode and the tip of a support needle and make one or more small welds to attach the wire to the tip at one or more points. After both ends of the wire are welded, increase the current and use the electrode to cut off the free ends of the wire. With experience, a hot wire sensor can be replaced in about 5 minutes.

Although one cannot repair hot film probes of the wedge type if they are burned out, minor cracks in the insulation coating can be repaired that would otherwise lead to eventual burnout. To repair a crack, clean the probe and paint the crack with lacquer, using a narrow artist's brush. Because lacquer lacks the mechanical strength of quartz, the probe should be inspected periodically, and more lacquer added if necessary. Cylindrical hot film probes can be repaired at the factory, and kits are available to allow a laboratory repair.

## 2.2 Probe fouling

Probe fouling may render a test program impossible to complete if contaminants cannot be removed from the fluid. If a contaminated fluid is used, some degree of probe fouling will reduce frequency response and sensitivity in an unpredictable way, and the probe will require repeated cleaning and recalibration for accurate measurements.

### Probe fouling in gases

Even laboratory air may be contaminated by dust, lint, flakes of dried wax from floors, or small droplets of oil from nearby machinery. Air in industrial environments may be heavily contaminated with oil droplets and residue from the products being manufactured. Outside air may also be contaminated, especially if measurements are required near roadways or airport runways. Probes used in recirculating wind tunnels can be fouled by oil from the propeller drive mechanism, whereas nonrecirculating wind tunnels using bottled gas are usually quite clean.

In air the fouling process may begin with the attraction by static electricity of a few oil and dirt particles to the sensor. As its surface becomes gradually covered with particles, the sensor diameter and roughness increase, and particles are more likely to become attached. The contaminants closest to the sensor are heated by it while being insulated by the layers above, and dissociation of the oil tends to cause the contaminants to harden into a more impervious layer. Particles that are damp with water may initially stick to the sensor and then be blown away as the water evaporates. Sensors used in temperature measurements are heated only slightly above ambient temperature, allowing damp particles that become attached to remain longer, which results in increased fouling of these sensors (Larsen and Busch, 1974). Sensors inclined at an angle to the mean velocity vector are less likely to become fouled, especially by long fibers.

Whereas fouled sensors usually have reduced velocity sensitivity, salt encrustation on sensors does not affect velocity sensitivity, because the thermal conductivity of the salt layer is much greater than the thermal conductivity of air (Schacher and Fairall, 1976).

### Probe fouling in liquids

Contamination of water occurs in many ways. Dust and dirt can drop from the air onto the surface of open tanks of water, and scale from the sides of the tank can be dislodged as well, although this is less severe when closed, nonreactive containers are used. If sufficient levels of light are present, the growth of one-celled animal and plant life can also be a problem.

In water measurements a sensor can be fouled by air bubbles (Rasmussen, 1967). Despite their size, bubbles influence the heat transfer between the sensor and the fluid. As bubbles grow and are swept away, the velocity sensitivity changes in an unpredictable way.

Bubbles on the sensor are best detected by visual inspection, and this means the sensor must be visible at all times. Preferably, one should not have to watch the probe through the undulating free surface of the water, although this may be unavoidable.

Air usually dissolves in water by contact at the air–water interface, and air dissolves faster if the water is moving. Although most waterborne air is dissolved and thus contained in the water as molecules of air dispersed between molecules of water, some small bubbles can occur, especially in the presence of surface disturbances. If the pressure of aerated water is decreased suddenly – for example, when a water faucet is opened and water at the pressure of the water main flows into a container at atmospheric pressure – the dissolved air can form into small bubbles (Rasmussen, 1967). Air bubbles can also form if aerated water is heated; an example is the heating of the water next to the sensor. Air bubbles carried along with the water sometimes develop an electrostatic charge that may cause the bubbles to be attracted to the sensor (Rasmussen, 1967). If the probe is dipped into the water, a bubble of air may be carried along to foul the sensor. According to Rasmussen (1967), this problem is greatest when the sensor is hot.

Bubbles can also form on the sensor by electrolysis; a crack or pinhole in the quartz insulation coating of a hot film sensor allows an electric current to pass from the sensor through the water to ground, generating hydrogen and oxygen gas bubbles that remain attached to the sensor.

The presence of an air bubble on a sensor causes a hot spot that overheats and prematurely ages the sensor at that point, because the thermal conductivity is much less for air than for water. If the bubble covers a pinhole or crack in the insulation coating, electrolytic etching can cause deterioration of the metal film, leading to early failure.

There are several ways to reduce the formation of gas bubbles. If the sensor temperature is low, there will be less opportunity for dissolved gases to be driven out of the liquid to form bubbles. This means the probe should be operated at as low an overheat ratio as possible while maintaining adequate velocity sensitivity. Keeping fluid velocity high has the double benefit of reducing the time that a particle of water is in the thermal field of the sensor and increasing the possibility that the liquid will sweep away any bubbles that form. If a probe is placed at an angle to the flow, there will be less likelihood of fouling by bubbles. Finally, since pinholes and cracks in the insulation coating of film probes lead to the formation of gas bubbles, these spots should be regularly inspected and repaired.

Before the development of insulated hot film probes, measurements were made in water, using uninsulated hot wire and hot film probes, although bubble-fouling problems were formidable. Stevens, Borden, and Strausser (1956) found that bubble formation was reduced by heating the sensor with an alternating current. Although they used the constant current mode of operation, a constant temperature anemometer supplying an alternating current to heat an uninsulated hot film probe for ocean measurements was developed by Grant, Stewart, and Moilliet (1962). Bubble fouling of uninsulated probes is also reduced if the sensor has low electrical resistance, because there will be less likelihood of a current path developing through the water to cause electrolysis. Reducing the resistance of a sensor does, however, reduce its sensitivity.

### Detection of fouled probes

Probe fouling is insidious because a probe may foul without your knowledge. There are only two ways to find out: inspect the probe visually, or recalibrate it to check for a change in sensitivity. Of the two, a recalibration is the only accurate method. To recalibrate, the probe is removed from the test fixture, put in a calibration device, and the full calibration procedure is repeated. If the calibration facility is designed to be part of the test setup, the probe can be calibrated while in place. If calibration over the entire velocity range is not possible, a check at one or two points may be sufficient.

With experience, one may be able to look at a probe and know when fouling has started to degrade performance. This experience is gained, however, only by regular inspection and recalibration of the probe. It may be possible to design the experiment so that measurements can continue while the probe is being inspected. In tests near a transparent wall, a microscope can be mounted on the other side for probe inspection and measurement of the sensor-to-wall distance. In the field there is less freedom. For example, one will be unaware of seaweed that fouls a probe attached to a body towed behind a ship. If the probe is brought to the surface for inspection or recalibration, the seaweed may dislodge, eliminating any chance of knowing whether or not the data should be discarded.

The detection of bubble-fouled probes used with liquids in the laboratory is not difficult if the probe can be observed, because the bubbles are easily seen with the unaided eye; but visual inspection in ocean or estuarine applications is virtually impossible.

Sometimes the presence of bubbles on the sensor can be detected by observing the output voltage signal from the anemometer. Bubble fouling is usually indicated by a sudden increase in the output signal, caused by a large bubble on the sensor being swept away. But this method detects only an advanced stage of fouling characterized by the growth of large bubbles rather than the gradual covering of the sensor by the small bubbles often encountered.

### Removal of contaminants from fluids

It is often better to either clean the fluid or to use clean fluid than to repeatedly clean the probe if fouling is a problem. Sometimes the fluid cannot be cleaned; for example, if a probe is used for air velocity measurements in an underground mineshaft, it will be in constant contact with damp, dusty, oily air, and any attempt to filter the air might disrupt the flow patterns under investigation. In the laboratory, clean fluids can be used.

A gas can be cleaned with filters or electrostatic precipitators. Filters can be installed in recirculating wind tunnels, and their use can be combined with regular cleaning of the tunnel walls. As an added precaution, the wind-tunnel test section should only be opened for short periods to prevent the entry of dust.

In-line filters can be used in recirculating water systems, and ordinary coffee filters work well. If microscopic examination of a filter shows only minor contamination after a period of use, the water can be considered clean (Delleur, Toebes, and Liu, 1968). For tanks of water, swimming pool pumps and filters can be used when tests are not in progress.

Noncorrosive tanks made of glass, plexiglass, or fiberglass do much to prevent the buildup of rust and dirt in the water. Covering these containers at all times, or at least when tests are not in progress, reduces contamination by dust falling on the water surface. If the covers are opaque, the tank will remain dark, inhibiting the growth of plant and animal life. If this does not overcome the problem, chemicals normally used to eliminate such growth in swimming pools can be added to the water.

There are several techniques available for removing bubbles and dissolved air from water. The simplest procedure is to allow the water to stand in containers for a period of time before use. Another method is to alternately heat and cool the water. If these are ineffective, the fluid pressure can be alternately increased and decreased or a vacuum can be applied while the fluid is vibrated in its container (Rasmussen, 1967).

### Cleaning probes

A probe can easily be cleaned by gently swirling it in a container of solvent. Then inspect it and recalibrate. Atlhough this procedure works well for both hot wire and hot film probes, the latter can be wiped with a soft artist's brush while in the solvent, as shown in Figure 2.1. A safer technique is to use an inexpensive ultrasonic agitator to vibrate the solvent while the probe is held stationary. Commonly used solvents are acetone and carbon tetrachloride. Hoffmeister (1972) used a mixture of chromic acid and sulfuric acid, and Jiminez, Martinez-Val, and Rebollo (1981) cleaned hot film probes by wiping them with a cotton swab dipped in acetic acid. After cleaning, rinse the probe in distilled water and allow it to dry. A novel reverse-flushing method of probe cleaning was developed by Grant, Stewart, and Moilliet (1962) for use on a probe attached to a submerged body towed behind a ship. It consisted of a cone-shaped housing behind the probe that would slide forward when the seawater pressure inside it was increased by a pump inside the towed body. This allowed a high pressure jet of seawater to be directed forward to back flush the probe for cleaning while submerged.

Bubble contamination on a hot film sensor can be removed by wiping with

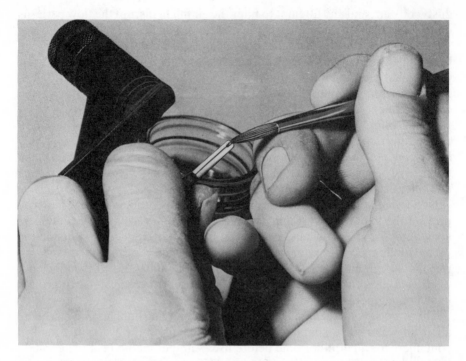

Figure 2.1. A hot film probe can be cleaned by dipping it in solvent and then wiping it with an artist's brush. Care must be taken not to bump the probe against the container or touch the sensor with the fingers while this is being done. Reprinted with permission from Dantec Elektronik.

a long-handled artist's brush, although one must be careful that air bubbles caught between the hairs of the brush are not transferred to the sensor.

### Using fouled probes

If neither probe nor fluid can be cleaned, then fouling cannot be eliminated – but accurate measurements may still be possible.

One method, developed by Richardson and McQuivey (1968) for turbulence measurements in contaminated water, is based on three assumptions: (1) The fouling material accumulates gradually with time; (2) the fouling material does not alter the sensor frequency response; (3) the calibration curve for a fouled probe is identical to that of a clean probe operated at a lower overheat ratio. This method also requires the use of a pitot tube for an independent measurement of mean velocity.

In this technique a clean probe is first calibrated in uncontaminated fluid at a variety of overheat ratios, and the results are graphed as a family of curves. Next, the probe is placed in the contaminated fluid, and turbulence measurements are taken. A pitot tube is used to measure mean velocity, and a comparison is made between this reading and that of the hot film probe;

a calibration curve is then chosen that applies for the type of fouling present. This is done by locating the output voltage on the ordinate of the family of curves and drawing a horizontal line. The pitot tube reading is located on the abscissa, and a vertical line is drawn. The intersection of these lines identifies the calibration curve, or, if the intersection falls between two curves, an interpolation can be made.

A method of taking measurements with a probe susceptible to bubble fouling was devised by Stevens, Borden, and Strausser (1956). They used an uninsulated hot wire probe in water attached to a hinged arm that was rotated down to immerse the probe for a quick velocity measurement and was then retracted. A microswitch was activated by the arm to turn on the probe heating current as the probe entered the water. In this way short-duration velocity measurements were made despite rapid accumulation of bubbles.

A second compensation method, the Sajben technique, will be discussed in the section describing measurements in mercury.

## 2.3 Traversing the probe

It is not unknown for a probe to be temporarily clamped into position at the beginning of an experiment for an operational check of the hot wire ane-mometer and then left there for the duration of the test without an adequate method for holding and traversing the probe. However, once the need for a probe traversing system is recognized, two requirements must be met; the traversing mechanism must move the probe while holding it rigidly, and the exact location of the sensor must be known at all times.

### Traversing mechanisms

A simple way to hold the probe is to C-clamp the probe support to V-blocks. Probe supports of various diameters can be accommodated, and the probe can be traversed by loosening the clamp and sliding the probe. Another device is a hole in a block, through which the probe support passes, with a setscrew to hold it in place.

An improvement is to advance the probe by means of screw threads; the screw rotation gives a direct measure of traverse distance. An example is a micrometer head attached to a handmade fixture. Such a device was used by Ling, Atabeck, Fry, Patel, and Janicki (1968) to traverse a hot film probe across arteries in blood flow measurements.

Traverse mechanisms can be purchased from optical equipment manu-facturers and modified to suit the application; they are available in one, two, or three degrees of freedom. If only a few centimeters of movement are needed, an *XY* traverse mechanism designed to position glass slides on a microscope can be modified. For greater stability a milling machine base can be used as a traverse mechanism. Although expensive even when pur-chased used, it can also be used as a traverse mechanism for a laser Doppler anemometer.

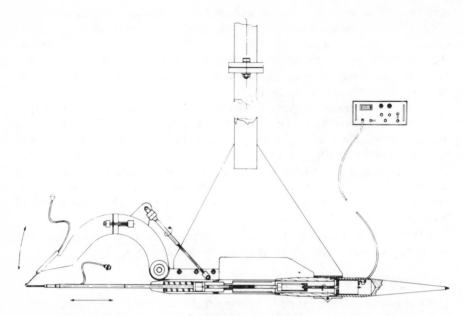

Figure 2.2. The probe support of this probe contains a small stepper motor used to traverse the sensor horizontally. The position of the second probe is adjusted manually. Reprinted with permission from P. Mestayer and P. Chaumbaud, Some limitations to measurements of turbulence micro-structure with hot and cold wires, *Bound. Layer Met.*, 16 (1979), 311–329.

Finally, traverse mechanisms with stepper motors can be used to automatically position the probe or move it steadily at a predetermined rate. Miller (1972), for example, designed a three-dimensional traversing mechanism for use in circular ducts that was capable of automatically traversing in concentric circles.

The stepper motor can also be housed inside the probe support, as the design illustrated in Figure 2.2 shows. In this figure the stepper motor advances and retracts one probe with a position accuracy of 5 μm; the location of the second probe is adjusted manually with an accuracy of 0.1 mm (Mestayer and Chambaud, 1979).

Stepper-motor-powered traverse mechanisms having three degrees of freedom are sold by suppliers of optical equipment. Excellent traverse mechanisms can also be purchased from manufacturers of hot wire anemometry. These devices are multispeed and operate under computer control to advance the probe in one or more directions across the flow.

A variation on the typical traversing mechanism is one designed to keep the sensor at a constant height above the surface of the water as waves pass by. A vertical wire acts as a capacitance transducer to sense wave height and control a motor by means of a feedback control circuit to maintain

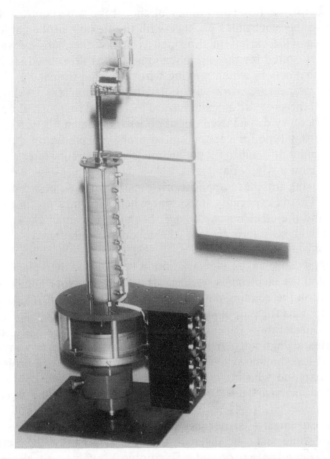

Figure 2.3. A vane-mounted array of hot wire probes for atmospheric measurements. Slip rings are located in the cylindrical housing at the base. Reprinted with permission from S. E. Larsen and N. E. Busch, Hot-wire measurements in the atmosphere. Part 1: Calibration and response characteristics, *DISA Info.*, 16 (1974), 15–33.

constant sensor-to-water distance (Shemdin, 1969 and Chang, Grove, Achley, and Plate, 1970).

### Rotating the probe

Sometimes the probe must be rotated through an angle of 360° or more. Figure 2.3 illustrates this requirement with a vane-mounted probe for atmospheric measurements. In this design a single-sensor probe is mounted at the top with a triple-sensor probe below it. A double-sensor probe for temperature measurements is located at the bottom. Slip rings are used for electrical connection to the six electronics packages required for this application. If only a few revolutions of the probe are required, it can be connected to the electronics package with the flexible probe cable.

There are a wide variety of slip rings available: the mercury wetted type is particularly suited for these applications. King (1914) used shallow mercury-filled channels in which moving wires traveled to make a type of slip ring for his pioneering work with hot wire anemometery. Although open containers of mercury are a health hazard in a closed room, shallow channels filled with mercury can be used out of doors. Larsen and Busch (1976) used slip rings of this type for vane-mounted probes located atop a high tower, but they found it difficult to fill the channels with mercury after a long climb up the tower on a windy day.

Each slip ring for these applications must have low, constant electrical resistance. This is especially important when slip rings are used between the probe and the electronics package, because any small change in resistance there will be interpreted by the feedback amplifier as a change in velocity.

If slip rings cause excessive variation in resistance when placed between the probe and the electronics package, the entire hot wire anemometer can be mounted on the rotating part (Kirchoff and Struziak, 1976), and the slip rings can be used for output signals and electrical power.

A final consideration is sensor stretching due to centrifugal force if the probe rotates at high angular velocity. This phenomenon was investigated by Hah and Lakshminarayana (1978) by covering a hot wire probe with its protective cap and whirling it at about 600 $gs$. The sensor cold resistance was found to increase by 1.8%, due, presumably, to sensor stretching.

### Measurement of sensor location

For some tests the distance between two sensor locations or the distance between the sensor and a fixed object, such as a wall, is needed. The second case has more potential for probe damage because an error can cause the probe to be traversed into the wall.

A simple way, often used in preliminary testing, to find the distance between two sensor locations is to hold a scale beside the probe support and measure by eye. However, a better method is to clamp a dial indicator to the traverse mechanism. Equally accurate is the micrometer head, described in the previous section, that acts both as a traversing mechanism and as a measuring device.

The distance between a sensor and a transparent outside wall can be measured with a microscope located on the other side. The microscope is focused on a mark on the inside surface of the glass to establish a reference position and then traversed forward a distance corresponding to the first measurement position. The probe is then moved toward the wall until it is in focus.

If the wall is opaque, the sensor location can be measured with a dial indicator or micrometer head if the exact position of the sensor relative to some point on the probe support is known. To do this remove the probe from the probe support and move the probe support forward until its end touches the wall in order to establish a reference distance. Measure the distance between the sensor and the other end of the probe body with an optical comparator or traveling-stage microscope and put the probe on the probe support. The location of the sensor is now known.

## 2.4 Vortex shedding from probes

Vortices shed from parts of the probe can cause errors if this effect influences the convective heat transfer at the sensor.

### The Karman vortex street

As the fluid passes over the parts of a probe, disturbances in the form of a regular pattern of eddies can occur downstream. This pattern, called the *Karman vortex street*, consists of two alternate rows of vortices. The frequency at which vortices are shed is expressed nondimensionally by the Strouhal number, Sn, defined as

$$\text{Sn} = \frac{f}{d}$$

where $f$ is the frequency at which vortices are shed from one side of the cylinder, and $d$ is the diameter of the cylinder. To illustrate the flow regimes over which vortices are shed, the Strouhal number is usually graphed versus the Reynolds number, Re, defined as

$$\text{Re} = \frac{Ud}{\nu}$$

where $\nu$ is the kinematic viscosity. Such a graph shows that vortices are only shed in the range Re < 50.

Eddies shed from the cylindrical parts of the probe can influence not only the sensor of this probe but the sensors of nearby probes as well.

Disturbances also occur when vortices are shed from the sensor itself. A graph of Strouhal number versus Reynolds number can be used to compare the shedding frequency expected with any periodic variation of the anemometer output voltage signal. To calculate the sensor Reynolds number, it is customary to evaluate fluid properties at the mean film temperature. Fabula (1968), however, obtained best results with fluid properties evaluated at the sensor temperature.

### The influence of vortex shedding on performance

Sinusoidal variations appear on the anemometer output voltage signal when vortices are shed from the parts of a probe, and this should be checked in each new application. For a more thorough check, traverse the wake of the probe with another probe from which no vortices are shed (Collis and Williams, 1959).

Vortex shedding is also influenced by dirt on the sensor. Fabula (1968) found that small amounts of fouling material significantly retard the onset of vortex formation. As a test, small spots of varnish were placed on the cylindrical sensor of hot film probes used in water, and, after this modification, velocities 25% higher than predicted by theory were required to cause vortex shedding.

In gases, vortices shed from the sensor may occur at a frequency far above the upper frequency limit of the anemometer. If so, the vortex shedding will not influence the anemometer output voltage. If the vortex shedding frequency is below the system upper frequency limit, then the output voltage can be low-pass filtered to remove the unwanted signal.

## 2.5 Directional sensitivity of probes

An advantage of hot wire anemometry is the ability of the probe to sense not only the speed of the flow but its direction as well, although two sensors are needed to measure each independently.

### The ideal cylindrical sensor in yaw, pitch, and roll    $\theta, \vec{v}$

To understand how a heated cylindrical sensor can be used to detect the angle between it and the velocity vector, the velocity vector can be separated into two components: a transverse component normal to the sensor, and a longitudinal component parallel to the sensor. Although both pass over the sensor, the transverse component is mainly responsible for sensor cooling.

The yaw angle, $\theta$, is defined for a standard hot wire probe as the angle between the velocity vector and the normal to the sensor, both of which lie in the plane of the support needles, as shown in Figure 2.4.

The effective velocity, $U_{eff}$, measured by the sensor is equal to the transverse component of velocity; that is,

$$U_{eff} = U_x$$

Because $U_x = U \cos \theta$, we have

$$U_{eff} = U \cos \theta$$

where $U$ is the magnitude of the mean velocity vector. This expression is called the *cosine law*, and tests by Champagne, Sleicher, and Wehrmann (1967) show that it gives excellent results if the sensor aspect ratio is greater than about 600. Because few realistic probe designs have such a large aspect ratio, the cosine law can only be used as an approximation.

The sensitivity of a probe to yaw is defined as

$$\text{Yaw sensitivity} = \frac{\partial E}{\partial \theta}$$

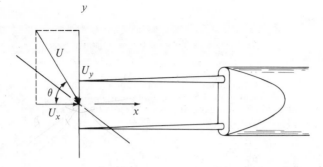

Figure 2.4. The yaw angle, θ, for a standard hot wire probe.

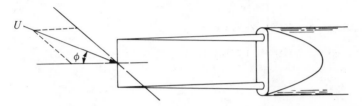

Figure 2.5. The pitch angle, ϕ, for a standard hot wire probe.

A probe is also sensitive to changes in pitch angle, ϕ, which is defined as the angle the probe body axis makes as the probe is rotated in such a way that the sensor does not change position. This is illustrated in Figure 2.5. Its pitch sensitivity is defined as

$$\text{Pitch sensitivity} = \frac{\partial E}{\partial \phi}$$

The roll angle, ψ, is the angle formed by the sensor as the probe is rotated about the probe body axis. Hot wire and cylindrical hot film probes (Figure 2.6) are not sensitive to roll, but some hot film probe designs do exhibit interesting roll characteristics. The roll sensitivity is defined as

$$\text{Roll sensitivity} = \frac{\partial E}{\partial \psi}$$

### Yaw- and pitch-angle relationships

For sensors having aspect ratios less than about 600, the cosine law is not accurate, but a simple modification to it was proposed by Hinze (1959, p. 103). The Hinze yaw-angle relationship is

$$U_{\text{eff}}^2 = U_x^2 + k^2 U_y^2 \tag{2.1}$$

where $k$ is the yaw factor, included to account for the additional cooling by the tangential component of velocity. This equation can be rewritten as

$$U_{\text{eff}}^2 = U^2(\cos^2 \theta + k^2 \sin^2 \theta) \tag{2.2}$$

Figure 2.6. The cylindrical hot film probe is an interesting compromise, using the support needles and probe body design of a hot wire probe and a sensor made by depositing a thin film of metal over a quartz fiber. Reprinted with permission from Dantec Elektronik.

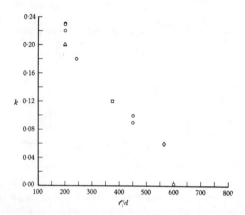

Figure 2.7. The variation in yaw factor for hot wire probes of different sensor aspect ratio. Reprinted with permission from F. H. Champagne, C. A. Sleicher, and O. H. Wehrmann, Turbulence measurements with inclined hot-wires. Part 1: Heat transfer experiments with inclined hot-wire, *J. Fluid Mech.*, 28 (1967), 153–176.

Webster (1962) determined the magnitude of the yaw factor to be $k = 0.20 \pm 0.01$ for most conventional hot wire probes.

Although the Hinze yaw-angle relationship is often used as the basis for theoretical investigations, its yaw factor is not constant but depends upon both sensor aspect ratio and yaw angle itself. The dependence of the yaw factor on sensor aspect ratio is shown in Figure 2.7, where the yaw factor decreases by a factor of about four as the aspect ratio increases from 200 to 600 (Champagne, Sleicher, and Wehrmann, 1967).

✳ The effect of yaw angle on yaw factor is shown in Figures 2.8 and 2.9. The results of two probe tests are shown, one a standard probe having support needles spaced 1.25 mm apart (Figure 2.8), and the other with support needles spaced 3 mm apart (Figure 2.9). Both sensors have identical aspect ratios. The standard probe shows more variation, with the yaw factor decreasing by a factor of four as the yaw angle is increased from 20° to 90°. The yaw factor of the probe having more widely spaced support needles decreases by a factor of somewhat less than three for the same angular change (Jorgensen, 1971), clearly showing that wider support needle separation reduces the dependence of the yaw factor on the yaw angle.

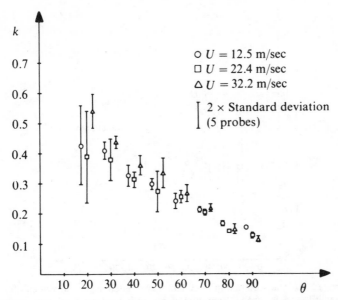

Figure 2.8. The variation in yaw factor with yaw angle and velocity for a standard hot wire probe. Reprinted with permission from F.E. Jorgensen, Directional sensitivity of wire and fibre-film probes, *DISA Info.*, 11 (1971), 31–37.

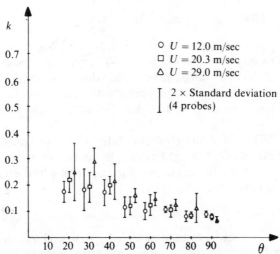

Figure 2.9. The variation in yaw factor with yaw angle and velocity for a hot wire probe having widely spaced support needles. Reprinted with permission from F. E. Jorgensen, Directional sensitivity of wire and fibre-film probes, *DISA Info.*, 11 (1971), 31–37.

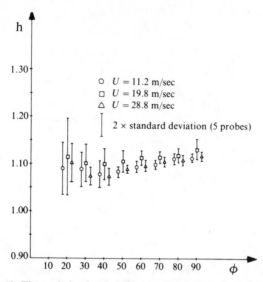

Figure 2.10. The variation in pitch factor with pitch angle and velocity for a standard hot wire probe. Reprinted with permission from F. E. Jorgensen, Directional sensitivity of wire and fibre-film probes, *DISA Info.*, 11 (1971), 31–37.

In an attempt to develop a yaw-angle relationship that fits the experimental data more accurately, Friehe and Schwarz (1968) developed the following yaw-angle relationship:

$$U_{\text{eff}} = U[1 - b(1 - \cos^{1/2} \theta)]$$

where $b$ is the yaw parameter, found by Friehe and Schwarz to be essentially constant for yaw angles of $-60° < \theta < +60°$, even though different for each probe. Although this yaw angle relationship does not simplify to the cosine law for large aspect ratios, it does represent the experimental yaw response data well and has been used by Drubka, Tan-atichat, and Nagib (1977) and Foss (1978).

Jorgensen (1971) added an extra term to the Hinze yaw-angle relationship to include the effect of pitching. Because the lateral component of velocity is assumed to cause additional cooling during pitching, this effect is included in the following equation:

$$U_{\text{eff}}^2 = U_x^2 + k^2 U_y^2 + h^2 U_z^2$$

where $k$ is the yaw factor, evaluated with $U_z = 0$, and $h$ is the pitch factor, evaluated with $U_y = 0$. Expanding this equation to include the yaw and pitch angles gives

$$U_{\text{eff}}^2 = U^2(\cos^2 \theta \cos^2 \phi + k^2 \sin^2 \theta \cos^2 \phi + h^2 \sin^2 \phi)$$

The pitch factor is not too dependent on pitch angle and sensor aspect ratio. Figure 2.10 illustrates how the pitch factor varies with pitch angle for

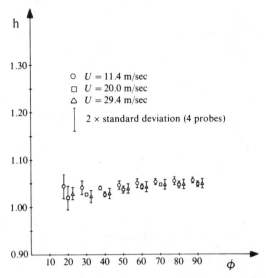

Figure 2.11. The variation in pitch factor with pitch angle and velocity for a hot wire probe having widely spaced support needles. Reprinted with permission from F. E. Jorgensen, Directional sensitivity of wire and fibre-film probes, *DISA Info.*, 11 (1971), 31–37.

a standard probe, and Figure 2.11 illustrates the response of a probe having widely spaced support needles. The pitch factor has values in the range $1.05 < h < 1.12$, depending upon probe type (Jorgensen, 1971), and only increases slightly with increasing pitch angle.

The response of a cylindrical sensor to pitch is due to the hydrodynamic effect of the fluid passing through the opening bounded by the sensor, support needles, and probe body when the pitch angle is increased. One theory attributes the pitch-angle effect to the additional cooling of the support needles because they are oriented broadside to the flow during pitching. There is, however, little evidence to suggest that the large variations experienced could be attributed to support needle cooling because only about 10% of the heat lost by a sensor is conducted away by the support needles (Dahm and Rasmussen, 1969). To test this hypothesis, Comte-Bellot, Strohl, and Alcaraz (1971) measured the pitch characteristics of a hot wire probe, coated with support needles with varnish to reduce convective cooling, and again measured the pitch characteristics. The pitch response was found to remain virtually unchanged.

### Measurement of angular response

Although manufacturers often provide yaw and pitch factor data for their probes, each probe should be calibrated to determine its yaw and pitch factors.

The probe body centerline is often used as a reference when positioning

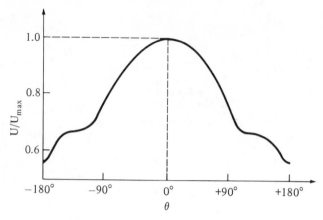

Figure 2.12. The yaw characteristics of a wedge hot film probe. Reprinted with permission from G. Mollenkopf, Measuring nonstationary periodical flow in the draft tube of a water-powered Francis model turbine, *DISA Info.*, 13 (1972), 11–15.

a sensor in a flow field, but poor quality control can result in sensors misaligned by 1° or 2°. Even if perfectly aligned when unheated, thermal expansion during use may cause the sensor to bow. Use of published yaw- and pitch-factor data with such a probe can cause serious errors.

Sensor angle can be measured with an optical comparator or a microscope having a protractor eyepiece. If these instruments are unavailable, the probe can be placed on the stage of an ordinary overhead projector and its image projected on a wall (Andreas, 1979). A large protractor can be used to measure the angles.

### Realistic probe designs in yaw, pitch, and roll

Some probe designs have unusual yaw, pitch, and roll characteristics. For example, yaw and pitch characteristics for a wedge film probe are shown in Figures 2.12 and 2.13. Notice the deviation from the cosine law in both cases and the extreme changes in sensitivity. Most disturbing is the pitch ambiguity near 0° shown in Figure 2.13, where a value of effective velocity could correspond to any one of six different pitch angles.

The spherical film probe (Figure 2.14), designed to be insensitive to either roll or yaw, is not completely free of variation in angular sensitivity. Its yaw and roll characteristics are shown in Figures 2.15 and 2.16.

## 2.6 Interference between probes

Sensors can interfere with one another when placed close together, but some applications require close placement. For example, when rakes of probes or multiple-sensor probes (Figure 2.17) are used, the sensor-to-sensor dis-

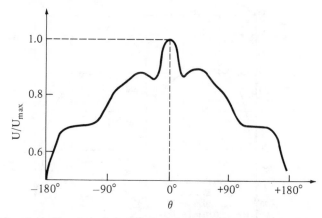

Figure 2.13. The pitch characteristics of a wedge hot film probe. Reprinted with permission from G. Mollenkopf, Measuring nonstationary periodical flow in the draft tube of a water-powered Francis model turbine, *DISA Info.*, 13 (1972), 11–15.

Figure 2.14. The spherical hot film probe is designed for minimum yaw and pitch sensitivity to allow it to measure the speed but be insensitive to the direction of the flow. Reprinted with permission from Dantec Elektronik.

tance is sometimes less than one sensor length, and interference between the sensors is possible.

There are two basic types of interference. One is hydrodynamic interference, an example of which is the change in the flow field that takes place close to a probe when a second probe is brought near. Another is thermal interference caused by the thermal wake of one probe passing over the sensor of another.

In a few applications, notably gas mixture concentration measurements and measurements with the pulsed wire anemometer, interference between the probes is desired, and probes are specifically designed to give the needed interference effect. For most multiple-sensor measurements, however, tests should be made to insure that interference does not occur.

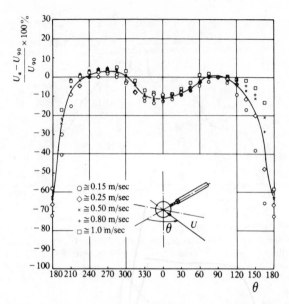

Figure 2.15. The yaw characteristics of a spherical hot film probe. Reprinted with permission from F. E. Jorgensen, An omnidirectional thin-film probe for indoor climate research, *DISA Info.*, 24 (1979), 24–29.

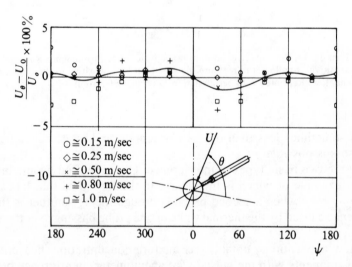

Figure 2.16. The roll characteristics of a spherical hot film probe. Reprinted with permission from F.E. Jorgensen, An omnidirectional thin-film probe for indoor climate research, *DISA Info.*, 24 (1979), 24–29.

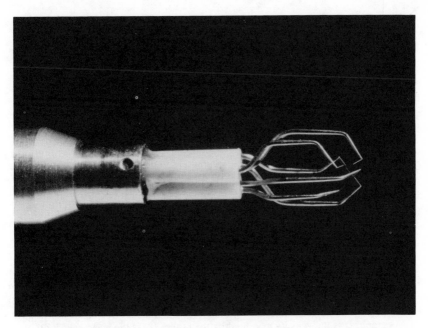

Figure 2.17. A triple-sensor hot wire probe having mutually orthogonal sensors. Reprinted with permission from P. Buchhave, Transducer techniques. *Proc. Dyn. Flow Conf.*, pp. 427–463. Skovlunde, Denmark.

**Hydrodynamic interference**

One would expect an object placed upstream from a probe to cause hydrodynamic interference, especially if the wake of the object passes over the probe. But tests show that probes placed near to each other, regardless of their relative position, can cause interference. Measurements of both mean velocity and turbulence intensity are influenced by hydrodynamic interference, and the effect of turbulence intensity has been found to be quite severe if the interfering probes are located near a wall.

The effect on mean velocity and turbulence intensity measurements when an unheated probe is placed directly upstream from a heated probe is shown by the "cold wake" curves of Figures 2.18 and 2.19, respectively. These curves show the unheated upstream probe to influence the mean velocity measurements of the downstream probe when the separation distance is as much as 18 mm. The turbulence intensity readings are influenced at a probe-to-probe distance of as much as 43 mm.

If the interfering object is downstream from the probe, the effect is much less. In tests by Comte-Bellot, Strohl, and Alcaraz (1971), cylindrical rods used to represent probe bodies were placed downstream from probes. The results, which clearly show that hydrodynamic interference occurs, are illustrated in Figure 2.20.

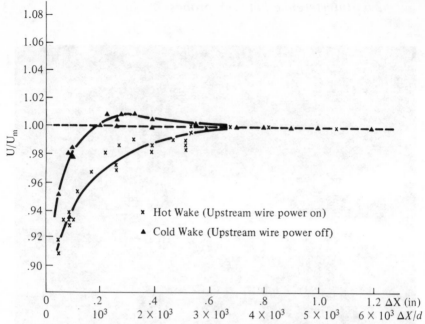

Figure 2.18. The variation in mean velocity measured by a hot wire probe placed in the hot and cold wakes of a second hot wire probe located upstream. In this graph, $U$ is the velocity measured by the probe, $U_m$ is the mean velocity of the fluid stream, $\Delta x$ is the sensor separation between the two probes, and $d$ is the sensor diameter. Reprinted with permission from N. W. M. Ko and P. O. A. L. Davies, Interference effect of hot wires, *IEEE Trans. Instr. Meas.*, 20 (1971), 76–78.

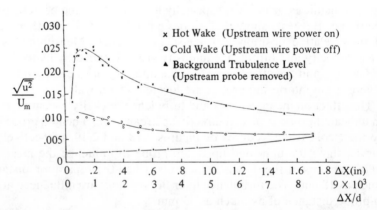

Figure 2.19. The variation in turbulence intensity measured by a hot wire probe placed in the hot and cold wakes of a second hot wire probe located upstream. In this graph, $u$ is the fluctuating component of velocity measured, $U_m$ is the mean velocity of the free stream, $\Delta x$ is the sensor separation between the two probes, and $d$ is the sensor diameter. Reprinted with permission from N. W. M. Ko and P. O. A. L. Davies, Interference effect on hot wires, *IEEE Trans. Instr. Meas.*, 20 (1971), 76–78.

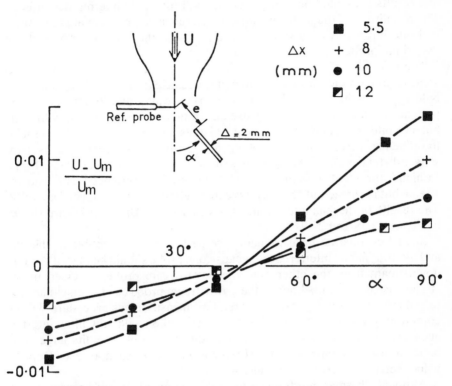

Figure 2.20. The response of a hot wire probe in the presence of a cylindrical rod located downstream. In this graph, $U$ is the measured velocity, $U_m$ is the mean velocity of the free stream, $\alpha$ is the angle between the cylindrical rod and the velocity vector, and $\Delta x$ is the separation distance between the sensor and the end of the cylindrical rod. Reprinted with permission from G. Comte-Bellot, A. Strohl, and E. Alcaraz, On aerodynamic disturbances caused by single hot-wire probes, *J. Appl. Mech.*, 38 (1971), 767–774.

If a probe is located near a wall and an object is brought near from above or from the side, serious turbulence intensity errors can result (Tritton, 1967). For example, the measurement of $\overline{u^2}$ was reduced as much as 35% when a disturbing probe was placed laterally nearby. Tritton also found this effect to be quite wall dependent, with errors of 5% noted when a measuring probe was 7 mm from the wall with the disturbing probe placed laterally, and errors of 30% when both probes were then lowered to a distance of 0.1 mm from the wall.

### Thermal interference

Although hydrodynamic interference between probes can be investigated by placing the disturbing probe nearby with its heating current turned off, thermal interference is more difficult to study because it is usually accompanied by hydrodynamic interference. In fact, all thermal interference test results presented here also include hydrodynamic interference effects. The magnitude of the hydrodynamic interference can be estimated in a thermal interference test by repeating the test with the heating current of the disturbing probe turned off.

The influence of the thermal wake of one probe on another depends to some degree on the turbulence intensity, because if the turbulence intensity is low there is little spreading of the thermal wake, and its effects are experienced a great distance downstream. In fact, in a flow in which the turbulence intensity is low, the thermal wake of a heated sensor has been found to influence the measurements of a probe located downstream at sensor-to-sensor distances as great as $4 \times 10^3$ sensor diameters (Ko and Davies, 1971). In highly turbulent flow the thermal wake is not experienced as far downstream but is spread out and may have more effect on probes placed laterally. The results of thermal interference tests by Ko and Davis (1971) are shown in Figures 2.18 and 2.19.

Spatial correlation measurements are particularly susceptible to thermal and hydrodynamic interference, because one probe should be placed directly downstream from the other, and the downstream probe traversed in the streamwise direction. Although the downstream probe can be offset laterally to reduce interference, correct correlation measurement procedure will be compromised, and spreading of the thermal wake could still allow interference unless the lateral spacing between the two probes is quite large. It is better to use a nonintrusive measurement technique, such as laser Doppler velocimetry, for the upstream measurement.

Although thermal interference is usually avoided, it can be used to advantage in some applications. For gas mixture concentration measurements, two sensors are located close together, one downstream from the other, to encourage thermal interference. The sensitivity of the array to gas mixture concentration is increased in this way. Another example is the probe used in pulsed wire anemometry that is designed to allow the heated wake of a suddenly heated wire to be convected downstream and over a second sensor. A typical pulsed wire probe is shown in Figure 2.21. Finally, wake-sensing anemometers, in which the thermal wake of one sensor is detected by a second in order to measure the angle at which the thermal wake leaves the upstream sensor, uses thermal interference to advantage.

## 2.7 Probe design

Design criteria for hot wire probes are often conflicting. For example, if an investigation of forced convection from a heated wire at an angle to the velocity vector is planned, end effects can be reduced if the wire is very long; but excess sagging of a long wire when heated makes the angle between the wire and the velocity vector uncertain. Researchers have separated design considerations and investigated them individually, and manufacturers use

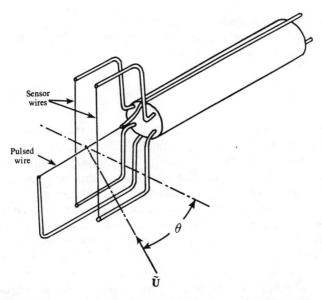

Figure 2.21. The probe for a pulsed wire anemometer has a large-diameter horizontal wire that is heated intermittently. It is flanked by two vertical wires that sense the "temperature spot" as it is convected past. Reprinted with permission from L. J. S. Bradbury and I. P. Castro, A pulsed-wire technique for measurements in highly turbulent flows, *J. Fluid Mech.*, 49 (1971) 657–691.

Table 2.1. *Optimum hot wire probe dimensions according to Strohl and Comte-Bellot (1973)*

| Design parameter | Dimensions (mm) |
| --- | --- |
| Support needle length | 20 |
| Support needle base diameter | 0.4 |
| Support needle tip diameter | 0.2 |
| Support needle spacing | 3 |
| Sensor length | 1 |
| Probe body diameter | 4 |

this information to produce general-purpose probes having wide enough appeal to be marketable.

Some probe optimization investigations have been reported in the technical literature, and others are proprietary to the manufacturers. Notable is the work of Gilmore (1967), Guitton and Patel (1969), Dahm and Rasmussen (1969), Comte-Bellot, Strohl, and Alcaraz (1971), Jerome, Guitton, and Patel (1971), and Strohl and Comte-Bellot (1973). Table 2.1 shows the conclusions by Strohl and Comte-Bellot (1973) for optimum hot wire probe dimensions.

**General sensor design considerations**

The resistance of a sensor depends upon both its dimensions and the resistivity of the sensor material. It is expressed as

$$R = \frac{\rho_r l}{A} \tag{2.3}$$

where $\rho_r$ is the coefficient of resistivity, defined as the resistance measured between opposite parallel faces of a sample of the material having unit length and unit cross-sectional area, $l$ is the length of the sensor, and $A$ is its cross-sectional area. The above equation shows that thin, long sensors have the greatest resistance, and high resistance is desired to maximize the signal-to-noise ratio and velocity sensitivity. Thus, for a given sensor diameter and length, a sensor material having high resistivity should be chosen. Alternatively, low sensor resistance is advantageous for an uninsulated sensor used in water to decrease the possibility of bubble formation by electrolysis.

Although resistivity values for the bulk material apply for a wire of the same material, this is seldom true for metal films used as sensors for hot film probes. Considerations such as plastic deformation of the film, impurities, trapped gases, oxides, and the fact that resistivity increases when the film thickness is less than the mean free path of the conduction electrons (Christensen, 1970) means that the film resistivity cannot be estimated. Instead, the sensor resistance must be measured after the hot film probe has been fabricated.

Probe resistance can be measured with an ohmmeter or by passing a very low-amplitude current through the sensor and using the anemometer Wheatstone bridge to measure the resistance. These techniques, however, measure a resistance that is the sum of the resistances of the sensor, support needles, probe leads, probe cable, and connectors. To find the resistance of the sensor alone, one can measure the resistance of a cable and a probe having no sensor and then subtract. This method, however, causes the resistance of the welded or soldered joint to be included with the sensor resistance.

The temperature coefficient of resistivity of a sensor material is used to express the relationship between the resistance and the temperature of the sensor. The temperature coefficient of resistivity, $\alpha$, is related to temperature and resistance by (Hinze, 1959, p. 78)

$$R_s = R_o[1 + \alpha(T_s - T_o) + \alpha_1(T_s - T_o)^2 + \cdots] \tag{2.4}$$

where $T_s$ and $R_s$ are the temperature and resistance of the heated sensor, and $T_o$ and $R_o$ are the temperature and resistance of the sensor at some reference temperature. The reference temperature is usually chosen to be 0°C, a temperature that can easily be reproduced in the laboratory with an ice bath. Values of the temperature coefficient of resistivity for typical wire sensor materials are (Hinze, 1959, p. 78)

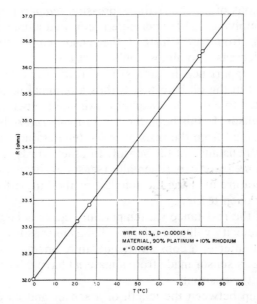

Figure 2.22. Resistance vs. temperature characteristics of a hot wire sensor made of 90% platinum and 10% rhodium. Note the linearity of this graph. Reprinted with permission from J. Laufer and R. McClellan, Measurement of heat transfer from fine wires in supersonic flows, *J. Fluid Mech.*, 1 (1956), 276–289.

Platinum:  $\alpha = 3.5 \times 10^{-3}(°C)^{-1}$

$\alpha_1 = -5.5 \times 10^{-7}(°C)^{-1}$

Tungsten:  $\alpha = 5.2 \times 10^{-3}(°C)^{-1}$

$\alpha_1 = 7.0 \times 10^{-7}(°C)^{-1}$

Because $\alpha \gg \alpha_1$ for typical sensor materials, the quadratic term is negligibly small, and eqn. 2.4 becomes

$$R_s = R_o[1 + \alpha (T_s - T_o)] \tag{2.5}$$

A typical calibration curve for the resistance vs. temperature characteristics of a 90% platinum, 10% rhodium sensor wire is shown in Figure 2.22. When the sensor is heated, its resistance will increase; the relationship is linear as well. The temperature coefficient of resistivity is seen to be represented by the slope of the resistance vs. temperature curve.

The resistance vs. temperature curve can be obtained by placing the probe in an oil bath or oven and slowly raising the temperature from ambient to a temperature higher than the operating temperature of the sensor. Place a thermometer or thermocouple beside the probe and measure temperature and resistance simultaneously. The resulting graph can be used to determine the temperature of sensors heated to their operating point.

In earlier sections of this book the expression *temperature of the sensor* has been used, implying that temperature is constant along the length of the sensor. This is far from true – instead, the sensor temperature is highest at the center and decreases to almost ambient at the ends. The oil bath test, however, heats support needles and sensor to the same temperature. The temperature used in the oil bath test is thus a nominal temperature and is not directly related to either the mean or maximum temperatures of an electrically heated sensor. In this book the *sensor temperature* is defined as the temperature obtained from the resistance vs. temperature curve and based on the operating resistance of the sensor. In other words, the sensor parameter easiest to measure experimentally (sensor resistance) is chosen and related to temperature by using the simplest method to perform experimentally (oil bath or oven test).

If the slope of the resistance vs. temperature curve is large, then a small change in fluid velocity will give a larger change in sensor resistance. Since the feedback amplifier of a constant temperature anemometer responds to sensor resistance, a sensor made from a material having high resistivity will have good velocity sensitivity.

The relationship between the length of a sensor and its diameter is important in hot wire anemometry. The sensor aspect ratio is defined as the ratio of the sensor length to its diameter:

$$\text{Aspect ratio} = \frac{l}{d}$$

The sensor aspect ratio should be in the range $200 < l/d < 400$ for best performance, according to Bruun (1971). A typical commercially manufactured hot wire probe, with a sensor length of 1 mm and a diameter of 5 $\mu$m, has an aspect ratio of 200.

Rules-of-thumb regarding sensor aspect ratio are based on a number of conflicting design specifications. The sensor should, for example, be as short as possible to reduce the likelihood of breakage due to shock, vibration, or impact. But a long sensor will lose, by conduction to the support needles, a smaller fraction of the total heat it generates, and this has three important consequences. First, the temperature profile on a long sensor will tend to be more uniform over its length, making the sensor temperature defined above closer to the average temperature of the sensor. Also, some measurements require calculation of the sensor Nusselt number, and although this can easily be done for a sensor of infinite length, a somewhat ambiguous correction is required for a short sensor having significant end losses. Finally, velocity sensitivity is improved if the sensor temperature is relatively high, and a long sensor has a significant portion of its length at high temperature.

Another consideration is thermal noise – that is, electronic noise generated by the sensor (Johnson and Llewellyn, 1934) due to the random movement of molecules, causing a variation in the charge between the two ends of the sensor. This effect is present in every wire and component in the anemometer and represents the minimum noise level possible for the system. Because thermal noise is about one order of magnitude less than the noise generated in the solid state amplifiers (Christiansen, 1970), it may be disregarded.

Sensors seldom melt when used in room air because an overheat ratio giving adequate velocity sensitivity will not bring the sensor temperature near its melting point. But if measurements are taken at high ambient temperature and an overheat ratio is chosen to give good velocity sensitivity, the sensor temperature may exceed the melting temperature of the sensor material. Even if the sensor does not reach its melting temperature, operation in air could cause oxidation, leading to rapid changes in sensor cold resistance, weakening, and eventual failure.

### Wire sensor design

Sensor diameter is an important design parameter for a hot wire sensor because sensor frequency response (as will be shown in a later chapter) is inversely proportional to the sensor mass, which has a stronger dependence upon diameter than length. Thus, the sensor frequency response and, consequently, the system frequency response can be improved by using a small-diameter sensor. In addition, the improved aspect ratio resulting from the use of thinner sensors reduces end effects.

When a wire sensor is heated, thermal expansion causes it to bow noticeably, making the angle between the sensor and the velocity vector ambiguous. If this probe is placed in turbulent flow, the velocity fluctuations can cause the bowed sensor to spin in circles or, more often, whip back and forth through an arc of 180° (Perry and Morrison, 1971b). Bowing can be reduced by pretensioning the wire during attachment to the support needles.

If a microscope cannot be used to observe sensor motion, a method developed by Perry and Morrison (1971b) might be useful, although it is less reliable. In this method a probe is calibrated at several different overheat ratios, and a discrepancy in the measurements at high overheat ratios may indicate that sensor motion is taking place.

The most commonly used materials for hot wire sensors are tungsten, platinum, and platinum alloys. Of these, tungsten is the most popular. Tungsten sensors are used unplated, with ends plated, completely plated, or in the form of Wollaston wire. Tungsten has a high tensile strength of 450,000 psi and is therefore an excellent choice for applications where a rugged sensor is required. It also has a high melting point and can be used in a nonoxidizing gas at a sensor temperature of 2000°C without melting (TSI catalog, 1978). A major limitation is its rather low oxidation temperature, and in air the oxidation temperature, rather than the melting point, sets the upper limit for sensor operating temperature at about 300°C (Disa Elektronik A/S probe catalog, 1980). Tungsten also cannot be soldered to the support needles, but it can be attached by spot welding.

The ends of tungsten wires can be plated with gold, nickel, or copper to eliminate some of the disadvantages of unplated tungsten sensors. Plating allows tungsten sensors to be soldered to the support needles. And since the plating material has high electrical conductivity, the wire ends will not be heated, reducing end losses and placing the active part of the sensor farther from the support needles to reduce hydrodynamic effects.

Tungsten wire plated with a thin layer of platinum is used by several probe manufacturers. It has, unfortunately, only a slightly higher oxidation temperature than unplated tungsten and must be attached by spot welding to the support needles.

The use of Wollaston wire allows a tungsten wire to be soldered to the support needles and is quite popular with researchers who make their own probes. The Wollaston process for manufacturing thin wire requires that a thick wire be given a heavy silver coating before it is passed through a die to bring it to correct size. The Wollaston wire is soft soldered in place on the support needles, and the silver outer layer is etched away by using a suitable acid. An example of a probe made using Wollaston wire is the X-array probe shown in Figure 2.25 (p. 45).

Platinum wire is another popular choice, especially by those who make their own probes. The tensile strength of unannealed platinum wire is 52,000 psi, which limits its use to lower-velocity flows. Its oxidation temperature is higher than that of tungsten wire, and it may be used in air at an operating temperature of about 1100°C (TSI Inc. catalog, 1978). Platinum sensors may be attached to support needles by either welding or soldering.

Of the alloys, platinum–iridium and platinum–rhodium are good choices for a wire sensor to be used in high-temperature air measurements. They have high oxidation temperature and high tensile strength and are capable of being operated at temperatures as high as 800°C (TSI Inc. catalog, 1978).

### Film sensor design

There are several methods of applying a metal film onto a quartz substrate. Of these, only cathodic sputtering and the paint-and-fire technique are used to make film probes for hot wire anemometry.

In the sputtering process the quartz substrate is placed in a vacuum chamber containing two electrodes. The substrate is placed near the negative electrode, which is made of the metal to be sputtered, and the chamber is filled with a gas such as argon at reduced pressure. A high voltage difference generated between the electrodes ionizes the gas, and positive ions from the gas bombard the negative electrode. The mechanical impact of these ions gradually disintegrates the negative electrode, throwing its molecules in all directions, and those reaching the quartz substrate are deposited. Because the sputtering material comes from all parts of a rather wide electrode, there is less likelihood that particles of dust will cast a "shadow" on the substrate to generate pinholes. The sputtered film is also uniform, and its thickness can be closely controlled.

In the paint-and-fire technique, thin layers of metallic paint are individually hand painted on the substrate by using a small artist's brush, and each layer is separately fired in a kiln. The film thicknesses are not as uniform as those produced by sputtering, but adequate hot film probes can be constructed.

The metal film should be protected by applying an insulation coating of quartz or lacquer by sputtering, using the paint-and-fire technique or painting with a brush. In any case the coating should have an adequate breakdown voltage, minimum thickness to minimize thermal lag through the insulation coating, and a coefficient of thermal expansion that is compatible with the metal film underneath to reduce cracking of the coating caused by heating and cooling of the metal film.

Besides the insulation coatings described above, additional coatings are sometimes applied to provide increased resistance to abrasion or, in the case of measurements in liquid metals, to provide a wetted surface. Abrasion by sand or other particles in the fluid can be so severe that, despite some loss in velocity sensitivity, extra coatings are needed. Two materials commonly used to prevent abrasion are lacquer, which is usually applied by dipping or painting, and teflon, which can be sputtered. Both are effective, and lacquer can be reapplied at the end of a day's testing if required.

Contaminants in liquid metals such as mercury adhere well to the unwetted surfaces of hot film probes. This can be reduced by sputtering gold or copper over the insulation coating to provide a wetted surface.

### Support needle design

Pitch-angle sensitivity occurs when the flow passes through the opening created by the sensor, support needles, and probe body. The flow is blocked by the probe body and accelerated as it passes between the support needles. To reduce these effects the support needles should be long, thin, tapered, and widely spaced.

Pitch sensitivity tests were conducted by Comte-Bellot, Strohl, and Alcaraz (1971), using a reference probe having very long and widely spaced support needles. A cylinder with the same diameter as a probe body was placed nearby, as shown in Figure 2.20. The effects of blocking and acceleration were found to cancel when the cylinder was at an angle of about 45° to the probe. Other tests by the same researchers also show that at distances of $\Delta x/d \geq 10$, where $\Delta x$ is the body-to-sensor distance, and $d$ is the probe body diameter, the effect of both blocking and acceleration become negligible. Dahm and Rasmussen (1969) found that their probes required separation distances of at least 2 mm between support needles to reduce pitch sensitivity to an acceptable level. A large separation distance need not result in a large aspect ratio for a sensor. In many commercial probes, for example, the sensors are about 1 mm long with about 1 mm plated on each side of the sensor to allow a separation distance between support needles of 3 mm.

Commonly used materials for support needles are steel, stainless steel, or phosphor bronze, and all have adequate strength. Eklund and Dobbins (1977) made support needles of constantan, which has a low coefficient of resist-

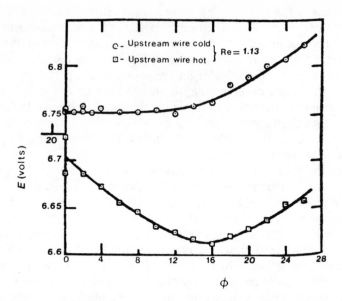

Figure 2.23. The pitch response of the downstream sensor of an X-array hot wire probe. Re = 1.13. Reprinted with permission from F. E. Jerome, D. E. Guitton, and R. P. Patel, Experimental studies of the thermal wake interference between closely spaced wires of a x-type hot-wire probe, *Aero. Quart.*, 22 (1971), 119–126.

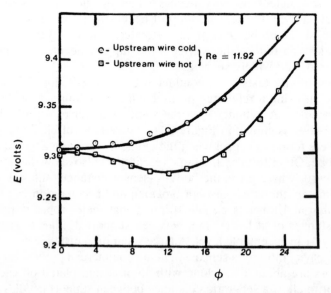

Figure 2.24. The pitch response of the downstream sensor of an X-array hot wire probe. Re = 11.92. Reprinted with permission from F. E. Jerome, D. E. Guitton, and R. P. Patel, Experimental studies of the thermal wake interference between closely spaced wires of a x-type hot-wire probe. *Aero. Quart.*, 22 (1971), 119–126.

ance, to reduce the effect of support-needle temperature variations on overall probe resistance. Support needles are sometimes gold plated to improve electrical conductivity. Although support needles are seldom insulated, their use in an electrically conductive fluid such as mercury requires a coating of varnish, lacquer, or epoxy paint.

### Probe body design

The probe body should be small in diameter, streamlined, and long enough for easy handling. If a probe is terminated with a cable for attachment to the electronics package, then it must be unclamped from the test fixture if repairs are needed. Although this design is used by those who make probes by hand, most manufacturers design short probe bodies terminated with waterproof electrical plugs to allow quick replacement of a damaged probe.

### Multiple-sensor probe design

The sensors of an X-array probe do not interfere with one another despite close spacing if the probe is not pitched. Jerome, Guitton, and Patel (1971) found that the sensor-to-sensor distance for an unpitched X-array probe could be as little as $0.161l$, where $l$ is the sensor length, with no adverse effect on performance. For an X-array probe in pitch, a sensor-to-sensor distance of $1.0\ l$ was needed to eliminate interference. A sensor separation of this magnitude does not compromise spatial resolution. These pitch errors were caused by the effect of the thermal wake from one sensor on the other. The results of these tests are illustrated in Figures 2.23 and 2.24. Notice that serious interference occurs when closely spaced sensors are pitched at low Reynolds numbers.

The design of the support needles and probe body of X-array probes was found by Strohl and Comte-Bellot (1973) to influence turbulence intensity measurements. An experimental X-array probe having long, widely spaced support needles was used to measure turbulence intensity in a test flow. The support needles and probe body of a commercially manufactured probe were then brought nearby, and errors of $-15\%$ were found in the measurement of turbulence intensity.

Misalignment of one or more of the sensors of an X-array probe by a few degrees can cause significant errors (Strohl and Comte-Bellot, 1973). A misalignment of one sensor by $1°$ was shown to cause errors in Reynolds stress of $1.5\%$. This error was seen to increase to $2.5\%$ if both sensors were out of alignment by $1°$.

### 2.8 Handmade hot wire probes

One cannot err in choosing a probe sold by a manufacturer if the design suits the application. They will usually make nonstandard probes for special applications, and although somewhat expensive and time consuming, the finished probe will be of excellent quality. A sketch showing all dimensions is required, and standard materials, support needles, and probe bodies should be specified where possible to reduce the final cost.

There are good reasons for making a probe by hand, and a certain justification for using only handmade probes. The parts for handmade probes are easily found and cost little. The time required to do the work is not great, especially after one or two probes have been made. Also, a person skilled in making probes can easily repair a broken one. Finally, some users experience great satisfaction in designing and building probes and enjoy the challenge of making each one better than the last.

### The decision to weld or solder the hot wire sensor

Both welding and soldering give strong, durable, and trouble-free joints. Each requires special tools and the development of manual skills for good results. In addition, either can be mastered by a person having only average mechanical aptitude and ability to do detailed work. Both require a holder for the wire sensor and the probe body, and these can easily be constructed. Finally, wire sensors can be welded or soldered to the support needles in about the same length of time – about five minutes for the experienced person.

Despite these similarities, each method has specific advantages. If welding is chosen, any wire material can be used, and the joint will withstand high ambient temperatures. Soldering is a skill that many possess, and no special equipment, other than a soldering iron, is required.

### Plating hot wire sensors

The ends of tungsten sensors are often plated to allow soldering to support needles. Copper is the usual plating material choice of those who make their own probes, although Spangenberg (1955) used nickel plating with good results. Some commercially manufactured probes have sensors with ends that are first plated with copper and then overplated with gold to increase the electrical conductivity of the plating.

To make a sensor having plated ends, one can either plate the ends of an unattached sensor and solder it to the support needles or plate the entire unattached sensor, solder it to the support needles, and etch the center portion with acid to expose the wire underneath.

To plate an unattached sensor, first make a wire holder from cardboard with a rectangular hole cut in its center (Davies, Tanner, and Day, 1968). Unwind the wire from its spool, wind the wire around the cardboard several times, and fasten with tape. Plate the wire bridging the hole by immersing it in the plating solution. To plate the ends of the wire, immerse only that part. For plating two segments of the sensor at one time, separate the solution container into two compartments and plate as before.

An acidified copper sulfate solution of 90% copper sulfate and 10% sulfuric acid can be used as an electrolyte with the addition of a small amount of wetting agent to reduce surface tension (Davies, Tanner, and Day, 1968). Use a potentiometer to control the plating current.

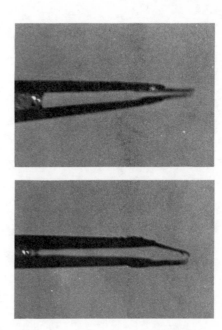

Figure 2.25. Drawings and photographs of top and side views of a handmade miniature X-array probe with Wollaston wires first attached to the support needles and then etched to expose the sensor wires. Notice that the unetched portion of the Wollaston wire acts as part of the support needles. Reprinted with permission from W. W. Willmarth and T. J. Bogar, Survey and new measurements of turbulent structure near a wall, *Phys. Fluids*, 20 (1977), S15–S16.

Davies, Tanner, and Day (1968) recommend a rather low plating current of 0.1 mA for the first 10 minutes to plate a thin, durable layer over the wire. Then the current is doubled, and plating is continued for an additional 20 or 30 minutes. At the end of this time the diameter of the sensor will increase from 5 $\mu$m to about 25 $\mu$m.

Plating can be removed from the center of the sensor by placing a small droplet of dilute sulfuric acid on the sensor to etch away the copper (Wills, 1962).

### Etching Wollaston wires

Wollaston wire can be attached to the support needles, after which a droplet of dilute nitric acid can be placed on the center to remove that part of the outer covering. Observe this process with magnification to obtain the correct sensor length. After etching, clean with distilled water. An example of a probe made by first soldering Wollaston wire to the support needles and then etching is the miniature X-array probe shown in Figure 2.25. The probe is so small that the unetched portion of the Wollaston wire forms part of the support needles. The wire lenths are 100 $\mu$m, and the wire diameter is 0.5 $\mu$m. The angles made by these sensors could not be set

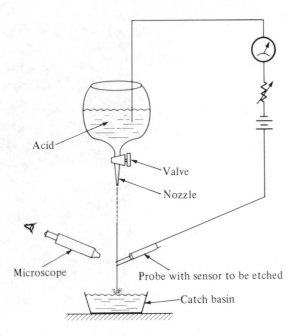

Figure 2.26. An apparatus using an acid jet for precise etching of Wollaston wires after attachment to the support needles. This process is monitored visually with a microscope. Reprinted with permission from R. Betchov and W. Welling, Some experiences regarding the nonlinearity of hot wires. NACA TM 1223, 1952.

precisely to 45°, so a calibration scheme was used to interpret the probe data.

If ease of handling is not important, etch Wollaston wire before attaching by connecting one terminal of a 1.5-V battery in series with a 10,000-$\Omega$ resistor to an end of the wire. The other battery terminal is connected to a container of dilute nitric acid. Dip the Wollaston wire into the acid, and use a current of about 0.1 mA to start the etching process (Sandborn, 1972, pp. 183–184).

A thin jet of acid can be used to accurately etch away the silver covering (Betchov and Welling, 1952), and etched lengths as short as 0.19 mm are possible. An apparatus for this purpose is shown in Figure 2.26. To use this technique, place the acid in a glass container having a small glass valve and nozzle at the bottom and adjust the valve to allow a thin column of acid to pass over the wire. Connect the acid, a mixture of 50% nitric acid and 50% distilled water, to a battery in series with a potentiometer and use a current of 5 to 20 mA. Watch the etching process and adjust the valve to gradually increase the length of the exposed sensor.

### Jigs for holding wires and probe bodies

Both sensor and probe body must be held securely when soldering or welding. With either method the wire sensor must lie flat against the tip

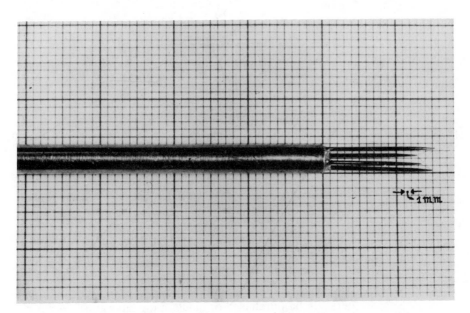

Figure 2.27. A handmade vorticity probe of the Kovasznay type having four support needles and four sensors. Reprinted with permission from E. G. Kastrinakis, J. M. Wallace, W. W. Willmarth, B. Ghorashi, and R. S. Brodkey, On the mechanism of bounded turbulent shear flows, in *Lecture Notes in Physics*, Vol. 75, pp. 175–189. Springer-Verlag, 1977.

of a support needle. A robust fixture should be used to hold the probe body because it receives the force from the soldering iron or welding tip. When soldering, you can use flux to hold the wire to the support needle during attachment.

There are a variety of ways for holding the wire sensor, but none are complicated. For example, tape the wire to the same type of cardboard holder used for etching Wollaston wires. Or attach a small weight, such as a piece of tape, to the end of the sensor wire hanging down from its spool; if the room air is still, the wire will hang motionless. This method was used by Willmarth (1978) to attach sensors inclined at 45° on a vorticity probe (a typical vorticity probe is shown in Figure 2.27). The probe body can be held with a table vise. Slide it along the workbench by hand until the support needles touch the sensor.

Sometimes wires are pretensioned to eliminate sag when heated. A small screw can be added to the probe holder to flex one support needle slightly inward while attaching the wire; when the screw is released, the wire will be held in tension (Champagne, Sleicher, and Wehrmann, 1967). Heat the wire while observing it under a microscope to determine if the pretensioning is adequate. A more accurate pretensioning scheme was developed by Lowell (1950), who used a fixture that pretensioned the wire while measuring its tensile force. Tensions of $0.414 \times 10^{10}$ dyn/cm$^2$ were found to be satisfac-

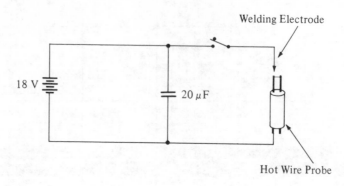

Figure 2.28. The circuit diagram for a spot welder used to attach hot wire sensors to the tips of support needles (Spangenberg, 1955).

tory. Best results were achieved when the wire was heated electrically to about 800°C during pretensioning.

### Welding wire sensors

Spot welders can be purchased or made in the laboratory. A simple circuit, developed by Spangenberg (1955), is shown in Figure 2.28. To weld a wire to a support needle, move the welding tip into position and momentarily close the switch to discharge the capacitor. The use of several welds along the wire on the tip of the support needle is good practice for secure attachment. Spangenberg found that silver or copper electrodes of about 0.13 mm diameter work best.

### Soldering wire sensors

Sensors are soldered to the support needles by first degreasing the sensor and support needles by dipping in turpentine, followed by rinsing in acetone and then distilled water (Delleur, Toebes, and Liu, 1968). Tin the support needles by melting a small piece of solder on the tip of a hot soldering iron, and transfer the solder to the tips of the support needles. The tips of the support needles should be evenly covered with a thin coat of solder. After allowing the support needles to cool, place a small amount of solder flux paste on the tip of a support needle. Bring the sensor wire near and push it into the flux to hold the wire against the tip. Place the tip of the soldering iron against the side of the support needle, but not touching the wire, to melt the solder that already coats the support needle. When the support needle cools, the wire will be attached (Sandborn, 1972, p. 187). Attach the sensor to the other support needle in the same way. For final cleaning, Delleur, Toebes, and Liu (1968) suggest making a small brush by binding together short lengths of sensor wire and using it to gently remove bits of solder and flux from the joint.

A small alcohol torch applies no force to the support needles and can be used instead of a soldering iron. The technique is identical to that described above, and the flame is directed to the side of the support needle below the joint to allow heat to be conducted to the sensor; the flame must not touch the joint at any time.

Doughman (1972) used the paint-and-fire technique to attach wires of 0.25 μm diameter by painting them in place with gold paint of a type to be described in the next section. Subsequent firing of the probe in a small oven at 540°C formed a gold bond that accurately joined the sensor to both support needles. One application of paint was found to be sufficient for 0.25-μm wires, but several layers were required for wires of larger diameter, and this was done by repeated painting and firing.

### Handmade probe bodies and support needles

Support needles are often made from ordinary sewing needles because they are of hardened steel, are plated to reduce corrosion, and are already tapered to a sharp point. They often need no modification; even the eye of the needle is retained to attach the cable leads. Small-diameter jewelers broaches (Blackwelder and Kaplan 1976, Fabris 1978, Foss 1978) can also be used; they are long and tapered and made of hardened steel, but they may require smoothing and shaping. Drill rod or wire, such as hardened steel music wire, stainless steel wire, or nickel wire can be used, but it usually requires shaping to improve streamlining.

Support needles can be shortened after attachment by cutting with wire cutters and shaping with a small grinding wheel or sharpening stone. Bend the support needles to shape by using two pairs of needle nose pliers, one held in each hand.

Support needles are usually circular in cross section, but other shapes have been used. Laufer and McClellan (1956) used stainless steel wedge-shaped supports in a supersonic probe design. In an alternative scheme developed by Mikulla and Horstman (1975), a ceramic wedge-shaped piece was cast between the support needles of a conventional probe to help support the sensor.

The probe body is often made of epoxy cast in a cylindrical shape with the probe cable potted into the end of the probe body. Small-diameter coaxial cables are available and are connected with one support needle attached to each conductor.

A variety of interesting hot wire probe designs is described in the technical literature. In an attempt to reduce capacitance and inductance in long lengths of probe cable, Wehrmann (1968) placed the feedback amplifier and Wheatstone bridge from the electronics package in the probe body and, in addition, piped liquid nitrogen around the amplifier for cooling to reduce noise. A detachable probe design is the "light bulb probe" of Hauptmann (1968). The glass envelope of 24-V pilot light bulb was carefully broken without damaging the filament by using modified pliers. The screw-in base with its coiled tungsten filament was used as a probe. It is screwed into a tube having a

Figure 2.29. A boundary-layer hot wire probe having a pin to protect against wall contact. This design allows only the sensor and the support needle tips to extend into the boundary layer to reduce flow disturbances. Reprinted with permission from TSI, Inc.

Figure 2.30. An inclined sensor hot wire probe. Reprinted with permission from TSI, Inc.

light bulb socket at one end. Although it violates every criterion for good hydrodynamic and thermal design, in applications where streamlining is not critical and cost and ease of replacement are important, this probe may find a use.

### Modification of commercially manufactured probes

One way to develop the ability to make a probe by hand is to begin by repairing the sensors of commercially manufactured probes. Do this by first inspecting the support needles under a microscope while filing and polishing any rough spots. Straighten the support needles, adjust their length by filing, and attach the sensor.

It is a small step from replacing broken sensors to modifying commercial probes for special applications, and there are good reasons for using them for your first design effort, because if commercially available probes have been used in the past, a collection of compatible probe supports and other accessories may be available.

The support-needle spacing of standard hot wire probes can be modified by giving each support needle a slight S-shaped bend or by bending each needle outward or inward slightly. A boundary layer probe (Figure 2.29) can be made by bending both support needles in the same direction (van Thinh, 1969). An inclined wire probe (Figure 2.30) can also be made by shortening one support needle.

### 2.9 Handmade hot film probes

Although commercially available hot film probes are made by a process that cannot be duplicated by hand, hot film probes can be made by a method developed by Ling (1955), the inventor of the hot film probe. These probes perform adequately and are only slightly more difficult to make than hot wire probes. They are usually fabricated when no commercially available design suits the application.

### The paint-and-fire technique

The paint-and-fire technique requires the individual firing of many coats of metallic paint to the substrate. Begin by applying to the glass substrate a thin layer of platinum, silver, or gold paint of a type used to decorate ceramics. Place the substrate in a cold oven and slowly raise the temperature to 640°C over a one-hour period. Hold that temperature for an additional half-hour, leaving the furnace door partially open at all times to ventilate the oven. If the oven is not ventilated or if the temperature is raised rapidly, the solvent in the paint will evaporate too quickly, leaving pinholes in the metal film (Bellhouse, Schultz, and Karatzas, 1966). Remove the substrate from the oven, allow it to cool, and repeat this process for the next coat of paint (Bellhouse and Bellhouse, 1968). Add layers of paint until a suitable electrical resistance is obtained for the metal film. Usually three to eight coats of paint are required (Christensen, 1970). In a modification of this technique, Seed and Thomas (1972) used a firing temperature of 600°C but of only 15 seconds duration. No oven was used; instead the substrate was placed inside a small coil of nichrome wire with its temperature controlled with a variable transformer.

After applying the metal film to the substrate, use epoxy glue to fasten the electrical leads to the substrate, and attach them to the film by either soldering with a small iron or by painting a connecting link between the wire and the film with electrically conductive paint of the type used on printed circuit boards (Bellhouse and Schultz, 1966).

A novel method of photomasking, developed by Seed (1969), allows metal films to be placed within 0.1 mm of each other. Begin by applying a metal film to a large section of substrate by using the paint-and-fire technique. Next, dip the probe in photoresist, a photosensitive paint, to cover the film and allow it to dry. Make a photomask by photographing a large black-and-white drawing of the shape of the film with a 35mm camera containing stripping film. The photosensitive material of the stripping film can be peeled off the paper backing and is flexible enough to be wrapped around the probe. Glue the mask in place on the probe and expose the probe to ultraviolet light. Remove the mask and immerse the probe in photoresist developer to expose the metal film that is not to be used. Etch away this film in an acid bath, leaving the metal film sensors in place, still covered with photoresist. Gold was chosen for the metal film because it could be etched easily.

Use a second photomask to plate the gold leads from the sensor with platinum so that electrical wires can be attached by soldering. Careful registration of the multiple photomasks is required, and this can be done by viewing the process under a low-power microscope. Then remove the photoresist covering the sensors and leads.

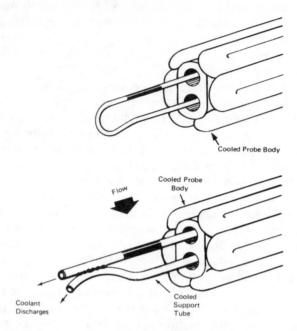

Figure 2.31. Two examples of cooled film probe designs; the top picture shows a probe that uses recirculating coolant, and the bottom picture shows a probe that allows the coolant to discharge into the fluid being measured. Although a thin quartz tube over which a thin, metal film has been deposited can be used as a sensor, the sensors in these designs are made either of ceramic or ceramic coated metal tubes over which the metal film is deposited. Cooling coils also cover the probe body. Reprinted with permission from TSI, Inc.

The sensor of a cooled sensor probe (Figure 2.31 shows a commercially manufactured example) can be made by dipping a small-diameter pyrex tube in platinum paint and firing. Then copper plate the ends of the tube and solder it to hollow support needles (Blackshear and Fingerson, 1962).

### Substrate construction

Glass is usually chosen as substrate material for handmade hot film probes, and rod and tubing can be purchased in a variety of diameters. Heat a glass rod, bend it to the desired configuration, and grind the end to form a cone, wedge or other shape. If electrical leads must pass through the substrate, use glass tubing instead. Close the end of the tubing by heating it over a Bunsen burner, and grind the closed end to the desired shape. Drill small holes in the end of the tube to allow attachment of the electrical leads to the metal film.

An easy way to polish the end of a glass rod after grinding is fire polishing: Place the ground end of the substrate momentarily in a Bunsen burner flame to slightly melt the outside surface of the glass and produce a smooth finish. Or the glass can be polished with a series of finer and finer abrasives.

Some of the smallest handmade hot film probes are made by applying a metal film to a bead of glass that is epoxy glued to a metal probe body. To make a bead, heat a glass rod at its middle with a Bunsen burner until it is soft and then pull each end of the rod with the hands to draw the glass out to a filament. After cooling, break the rod in half where the filament is thinnest. Heat the end of one filament to form a small bead of molten glass and allow it to cool. Break the bead away, leaving a small stalk of glass filament intact to support the bead when the metal film is applied. Then remove the stalk or leave it as a support when the bead is glued into the probe body. For a flattened bead, press the molten droplet against a flat piece of stainless steel. This flat surface can be improved by polishing (Bellhouse and Bellhouse, 1968).

### Application of insulation coatings
Insulation coatings can most easily be applied by painting or dipping. One method is to paint a thin coat of lacquer or epoxy resin over the metal film. To coat the sensor with epoxy resin, dip the end of the probe in thinned epoxy resin to give a coating about 10–15 $\mu$m thick. These coatings were found by Seed and Thomas (1972) to have an electrical resistance of about 25 M$\Omega$

Coatings can also be applied by a paint-and-fire technique developed by Vidal and Golian (1967). Cover the metal film by painting with a mixture by weight of 42% mother of pearl, 16% silicon resonate, and 42% gold oil thinner. After firing, the coating will be a combination of titanium dioxide and silicon dioxide. Aggarwal (1974) used this mixture and separately fired two coats at an oven temperature of 660°C. As added protection the edges of the sensor were painted with a single coat of varnish and baked in an oven at 150°C for five hours.

### Typical handmade hot film probes
Bellhouse and Bellhouse (1968) made a blood flow probe with two parallel sensors on a glass bead glued into a cutout in the side of a hypodermic needle that had previously been bent at right angles. Nerum, Seed, and Wood (1972) developed a three-sensor probe of the same configuration (see also Seed and Wood, 1970a). To construct these probes, the hypodermic needle was first bent at right angles, and the side near the tip was ground away. A glass bead with sensors and electrical leads in place was epoxied into the cutout with leads, as many as six in number, passing down the length of the tubing. The tip of the hypodermic needle remains intact for easy insertion of the probe through the wall of a blood vessel.

An interesting, flat-surface film sensor was built into the tip of a small 0.8-mm-diameter hypodermic needle (Ling, Atabeck, Fry, Patel, and Janicki, 1968) to measure shear stress at a blood vessel wall. The probe is inserted through the wall of the blood vessel and angled to be flush with the wall. For this design one lead passes through the inside of the hypodermic needle, and the needle itself acts as the second lead.

A ring-shaped sensor was developed for a flat-surface probe by Aggarwal (1978) to eliminate the directional sensitivity common to such probes. To make it, a thin glass tube was heated at the end and pressed against a surface to flatten it, almost closing off the end of the tube except for a small hole in the center. The end was lapped flat with diamond paste and dipped in platinum paint thinned to a 1:7 ratio by volume. The thin ring-shaped coating on the end of the tube was then fired at three different temperatures. Eight to 10 coats of platinum paint were applied in this manner. A copper wire was placed in the tube, and its end was glued to the center of the film with silver epoxy to hold the wire in place and connect it electrically to the film. The epoxy was cured by baking the probe in an oven at 130°C for one hour. the inside of the tube was filled with epoxy resin to pot the wire in place. This glass substrate was glued into the end of a hypodermic needle with silver epoxy, which also connected the platinum film electrically to the hypodermic needle. The second wire was soldered directly to the inside of the hypodermic needle tubing.

# 3  HEAT TRANSFER FROM SENSORS

Knowledge of a wide variety of disciplines, in addition to fluid mechanics, is required for a complete understanding of the hot wire anemometer; of these, thermodynamics, electronics, and automatic control theory are most important. In this chapter the thermodynamics of the probe will be discussed; this information will be needed to understand the operation of the electronic package to be discussed in a later chapter.

## 3.1 The heat balance for a hot wire sensor

Heat is lost from a hot wire sensor not only by convection to the fluid but also by conduction to the support needles. For typical sensor aspect ratios, this effect is seldom negligible, and for some measurements an accurate estimate of the fraction of heat lost in this way is necessary. Heat also leaves the sensor by radiation to cooler surroundings, although this effect is small and often neglected in calculations. Finally, heat may be stored in the sensor itself, and this effect is not negligible, even for thin wires having small mass. In fact, heat storage in the sensor limits the frequency response of the simplest type of constant current anemometer, making it unsuitable for turbulence measurements.

To analyze the energy balance for the hot wire sensor, we look at a differential element composed of a small length of the sensor, as shown in Figure 3.1. This differential element has length $dx$ and cross-sectional area $A$. The length of the sensor is $l$. Because the temperature profile of a sensor aligned normal to a uniform flow is symmetrical about the center of the sensor, the origin of the coordinate system is located there. Heat is generated electrically in this small differential element and dissipated by convection to the fluid, by radiation to the surroundings, and by conduction to the parts of the wire to which the differential element is attached. In addition, there is heat storage in the differential element of wire. These effects are expressed mathematically in an analysis adapted in part from Davies and Fisher (1964).

The rate of heat transfer into the left end of the differential element is

$$\text{Conduction heat transfer rate in left end} = -k_s A \left.\frac{\partial T_s}{\partial x}\right|_x$$

where $k_s$ is the coefficient of thermal conductivity for the sensor material,

55

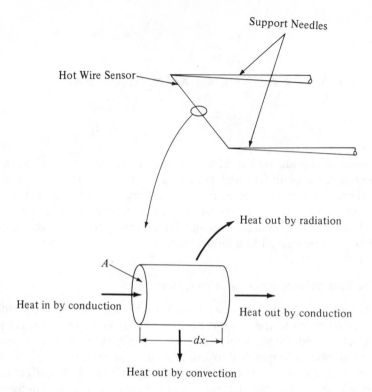

Figure 3.1. The differential element for a hot wire sensor.

and $x$ is measured along the sensor. The rate at which heat is conducted out of the right end of the differential element is

$$\text{Conduction heat transfer rate out right end} = -k_sA \left. \frac{\partial T_s}{\partial x} \right|_{x+dx}$$

A Taylor series expansion can be used to modify the above expression to become

Conduction heat transfer rate out right end

$$= -k_sA \frac{\partial T_s}{\partial x} - A \frac{\partial}{\partial x} \left( k_s \frac{\partial T_s}{\partial x} \right) dx$$

The heat generation rate is

$$\text{Heat generation rate} = \frac{I^2 \rho_r}{A} dx$$

where $I$ is the heating current, and $\rho_r$ is the resistivity of the sensor material.

The heat storage rate in the differential element is

$$\text{Heat storage rate} = \rho c A \frac{\partial T_s}{\partial t} dx$$

where $\rho$ is the density of the sensor material, $c$ is its specific heat, and $t$ is time. The heat rate out of the differential element by convection to the fluid is

$$\text{Convection heat transfer rate out} = \pi d h (T_s - T_f) dx \qquad (3.1)$$

where $d$ is the sensor diameter, $h$ is the coefficient of convective heat transfer, and $T_f$ is the temperature of the fluid. The heat transfer out of the differential element by radiation is

$$\text{Radiation heat transfer rate out} = \pi d \sigma \epsilon (T_s^4 - T_{\text{sur}}^4) dx \qquad (3.2)$$

where $\sigma$ is the Stefan–Boltzmann constant, $T_{\text{sur}}$ is the temperature of the surroundings, and $\epsilon$ is the emissivity of the sensor.

An energy balance yields the following differential equation for the heat balance in a hot wire sensor:

$$A \frac{\partial}{\partial x} \left( k_s \frac{\partial T_s}{\partial x} \right) + \frac{I^2 \rho_r}{A} - \rho c A \frac{\partial T_s}{\partial t}$$

$$- \pi d h (T_s - T_f) - \pi d \sigma \epsilon (T_s^4 - T_{\text{sur}}^4) = 0 \qquad (3.3)$$

This differential equation will be used later to find the equation for the temperature profile on the sensor as well as the average temperature and frequency response characteristics of the sensor.

## 3.2 Hot wire sensor temperature profiles

The temperature profile of a hot wire sensor is not constant but is greatest at the center and decreases to almost ambient temperature at its ends. In the following sections we define an expression for the shape of the hot wire sensor temperature profile and compare it to measurements of the temperature profile. An expression for the mean sensor temperature will also be derived.

### The shape of the hot wire sensor temperature profile

In order to derive the equation for the shape of the temperature profile on a heated cylindrical sensor, the differential equation for the heat balance of a wire sensor (eqn. 3.3) is used in an analysis that follows Davies and Fisher (1964).

If radiative heat loss is assumed to be negligible, sensor temperature assumed to not vary with time, and thermal conductivity assumed to be constant, then eqn. 3.3 becomes

$$Ak_s \frac{\partial^2 T_s}{\partial x^2} + \frac{I^2 \rho_r}{A} - \pi dh(T_s - T_f) = 0$$

If we use eqns. 2.3 and 2.5 this equation becomes

$$\frac{d^2 T_s}{dx^2} + \frac{\alpha I^2 R_0 - \pi dhl}{Ak_s l}(T_s - T_f) = -\frac{I^2 R_f}{Ak_s l}$$

Because the fluid temperature is constant along the sensor,

$$\frac{d^2 \theta}{dx^2} + K_1 \theta = K_2$$

where

$$\theta = T_s - T_f$$

and

$$K_1 = \frac{\alpha I^2 R_0 - \pi dhl}{Ak_s l} \tag{3.4}$$

$$K_2 = -\frac{I^2 R_f}{Ak_s l} \tag{3.5}$$

One boundary condition for this differential equation is provided by the intuitive notion that the temperature profile is maximum at the midpoint of the sensor; experimental verification of this characteristic of the temperature profile will be discussed next. This boundary condition is

$$\frac{d\theta}{dx} = 0 \quad \text{at} \quad x = 0$$

A second boundary condition is based on the assumption that the ends of the sensor and the support needles are at ambient temperature in uniform flow, or

$$\theta = 0 \quad \text{at} \quad x = \pm l/2$$

The constant $K_1$ may be either positive or negative depending upon the magnitude of $h$, but in practice it will be large enough so that $K_1$ is negative, and the solution becomes

$$T_s = \frac{K_2}{|K_1|}\left[\frac{\cosh(\sqrt{|K_1|}\,x)}{\cosh(\sqrt{|K_1|}\,l/2)} - 1\right] + T_f \tag{3.6}$$

This equation represents the temperature profile on a hot wire sensor. It is shown as the solid line in the next three figures.

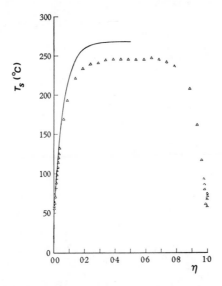

Figure 3.2. The temperature distribution along a hot wire sensor having an aspect ratio of 400. The nondimensional sensor length is $\eta$. Reprinted with permission from F. H. Champagne, C. A. Sleicher, and O. H. Wehrmann, Turbulence measurements with inclined hot-wires. Part 1: Heat transfer experiments with inclined hot-wire, *J. Fluid Mech.*, 28 (1967), 153–176.

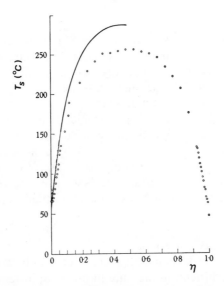

Figure 3.3. The temperature distribution along a hot wire sensor having an aspect ratio of 202. The nondimensional sensor length is $\eta$. Reprinted with permission from F. H. Champagne, C. A. Sleicher, and O. H. Wehrmann, Turbulence measurements with inclined hot-wires. Part 1: Heat transfer experiments with inclined hot-wire, *J. Fluid Mech.*, 28 (1967), 153–176.

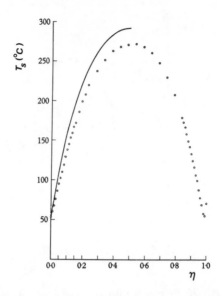

Figure 3.4. The temperature distribution along a hot wire sensor having an aspect ratio of 99. The nondimensional sensor length is $\eta$. Reprinted with permission from F. H. Champagne, C. A. Sleicher, and O. H. Wehrmann, Turbulence measurements with inclined hot wires. Part 1: Heat transfer experiments with inclined hot wire. *J. Fluid Mech.*, 28 (1967), 153–176.

The actual temperature distribution along a hot wire sensor was measured by Champagne, Sleicher, and Wehrmann (1967), using an infrared microscope, and the results are shown in Figures 3.2–3.4 for the case of sensors having aspect ratios of 400, 202, and 99, respectively.

Notice that for an aspect ratio of 99 the temperature is not constant anywhere along the sensor. Also, the sensor ends are seen to be almost at ambient temperature even though the center of the sensor is quite hot. The low end temperatures are due to heat loss by conduction to the support needles. Similar tests by Gessner and Moller (1971) for a hot wire sensor in shear flow give the temperature profiles shown in Figure 3.5. Notice that the temperature of the sensor ends remains virtually unchanged in shear flow.

### The mean sensor temperature

The mean sensor temperature may seem to be a useful parameter, but it is extremely difficult to measure in practice; instead, a "sensor temperature" found by using the oil bath technique is used to characterize the temperature of the sensor. But if the mean sensor temperature is needed, it can be calculated either by graphical integration of an experimentally determined sensor temperature profile or by integration of the equation for the theoretical temperature distribution.

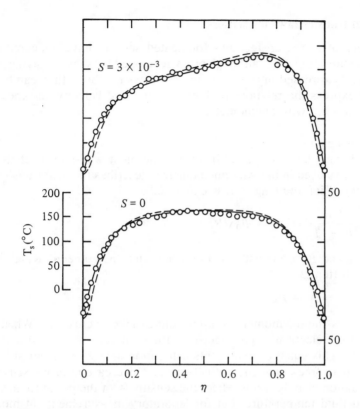

Figure 3.5. Influence of shear flow on the temperature profile for a hot wire sensor. The upper temperature profile is for shear flow, and the lower one is for uniform flow. The nondimensional sensor length is $\eta$. Reprinted with permission from F. B. Gessner and G. L. Moller, Response behavior of hot wires in shear flows, *J. Fluid Mech.*, 47 (1971), 449–468.

For a mathematical expression of the mean sensor temperature, eqn. 3.6 can be substituted into the following expression for the mean sensor temperature

$$\overline{T}_s = \frac{1}{l} \int_{-l/2}^{+l/2} T_s \, dx$$

to give

$$\overline{T}_s = \frac{K_2}{K_1} \left[ \frac{\tanh(\sqrt{|K_1|} \, l/2)}{\sqrt{|K_1|} \, l/2} - 1 \right] + T_f$$

Using this equation for an actual sensor requires calculating the constants $K_1$ and $K_2$; a method for doing this is described in Section 3.4.

### 3.3 Heat transfer laws for sensors

A variety of heat transfer laws for heated sensors have appeared in the technical literature, and, although often developed for sensors having infinite length and expressed in terms of dimensionless numbers, they can be modified to express the relationship between the fluid velocity and the output voltage of a hot wire anemometer.

#### King's law

King's law (King, 1914), by far the most well-known of the heat transfer laws used in hot wire anemometry, describes the heat transfer from a cylinder of infinite length. It is expressed as

$$\frac{I^2 R_s}{T_s - T_f} = A_0 + B_0 \sqrt{Re}$$

where $A_0$ and $B_0$ are constants. For a hot wire anemometer, King's law is usually written as

$$E_b^2 = A + BU^n \tag{3.7}$$

where $E_b$ is the anemometer output voltage taken across the Wheatstone bridge in the electronics package, $U$ is the fluid velocity, and $A$ and $B$ are constants. This equation represents calibration data well except at low velocity, where free convection effects cause a decrease in probe sensitivity. This equation also does not reflect the sensitivity of the probe to mass flow rate and fluid temperature, but for laboratory measurements at moderate velocities it is adequate.

King found the magnitude of the exponent $n$ to be 0.5, but subsequent experiments by Perry and Morrison (1971a) and Elsner and Gundlach (1973) show this exponent to vary from 0.45 to 0.5, depending upon velocity.

The addition of a third term to King's law (Siddall and Davies, 1972) allows its use over a larger velocity range. The equation is

$$E_b^2 = A + BU^{0.5} + CU \qquad \propto$$

where $A = 1.273$, $B = 0.860$, and $C = -0.017$. This "extended power" law gives excellent results for velocities as high as 160 m/s (Bremhorst and Gilmore, 1976).

#### Other heat transfer laws

Since the time when King mounted long, thin, heated wires on a rotating arm to verify his theoretical heat transfer law for infinite circular cylinders, others have experimented with high-aspect-ratio heated wires to find improved heat transfer laws. It is common for these results to be expressed in terms of the Nusselt number, Nu, defined as

$$Nu = \frac{hx}{k_f} \tag{3.8}$$

Table 3.1. *Values of the constants for the law of Collis and Williams for eqn. 3.9*

| Constants | Reynolds number range | |
|:---:|:---:|:---:|
| | $0.02 < Re < 44$ | $44 < Re < 140$ |
| $A$ | 0.24 | 0 |
| $B$ | 0.56 | 0.48 |
| $n$ | 0.45 | 0.51 |

*Source:* Collis and Williams (1959).

where $h$ is the convective heat transfer coefficient, $x$ is a characteristic length (usually the diameter of the cylinder), and $k_f$ is the thermal conductivity of the fluid evaluated at the mean film temperature, defined after eqn. 3.9.

An often-referenced heat transfer law is the result of experiments by Collis and Williams (1959), who found the heat transfer law for an infinite circular cylinder to be

$$\text{Nu} \left(\frac{T_m}{T_f}\right)^{-0.17} = A_1 + B_1 \text{Re}^{n_1} \tag{3.9}$$

where $T_m$ is the mean film temperature, defined as

$$T_m = \frac{\overline{T}_s + T_f}{2}$$

and the overbar denotes mean quantities. All fluid properties are evaluated at the mean film temperature. The constants in the law of Collis and Williams depend upon the Reynolds number, and Table 3.1 gives their values.

McAdams (1954, p. 260), in an attempt to find a heat transfer law for infinitely long, heated cylinders by collecting the experimental data from a variety of researchers, developed the following expression:

$$\text{Nu} = A_2 + B_2 \text{Re}^{n_2}$$

Another well-known heat transfer law for circular cylinders of infinite length is Kramer's law (Hinze, 1959, p. 76) expressed as

$$\text{Nu} = 0.42 \text{Pr}^{0.20} + 0.57 \text{Pr}^{0.33} \text{Re}^{0.50} \tag{3.10}$$

where Pr is the Prandtl number, defined as

$$\text{Pr} = \frac{\nu}{\alpha}$$

where $\nu$ is the kinematic viscosity, and $\alpha$ is the thermal diffusivity. Kramer's law is valid over the Reynolds number range of $0.1 < Re < 10,000$.

Another heat transfer law for cylinders is one developed by Van der Hegge Zijnen (1956). It is

$$Nu = 0.38Pr^{0.2} + (0.56Re^{0.5} + 0.001Re)Pr^{0.333}$$

The Collis and Williams law can be used for hot wire probes having finite aspect ratio by replacing the temperature ratio with the overheat ratio as

$$Nu(a_1 - 1)^{0.16} = A + B(Re)^{0.45}$$

according to Koch and Gartshore (1972).

These heat transfer relationships for hot wire sensors can be used if the Nusselt number can be expressed in terms of wire and fluid parameters, and this is easily done for sensors of infinite length. Assume the sensor temperature profile to be flat; this is equivalent to assuming conduction losses at the end of the sensor to be small compared to convection losses to the fluid. Also assume radiation losses to be negligible. Then the heat balance equation for a hot wire sensor (eqn. 3.3) becomes

$$\frac{I^2 \rho_r}{A} = \pi dh(T_s - T_f)$$

Substituting this equation into eqn. 3.8 to eliminate $h$ and using eqn. 2.3, we see that the Nusselt number for a cylinder of infinite length is

$$Nu = \frac{I^2 R_s}{\pi l k_f (T_s - T_f)} \tag{3.11}$$

Using the definition of temperature coefficient of resistance (eqn. 2.5), we then get

$$Nu = \frac{I^2 \alpha R_s R_f}{\pi l k_f (R_s - R_f)[1 + \alpha(T_f - T_0)]} \tag{3.12}$$

Notice that for a sensor operated in the constant temperature mode, all terms in the Nusselt number are constant except the sensor heating current. This equation can be used to approximate the Nusselt number of a hot wire sensor of finite length if its aspect ratio is large.

The disadvantage of using the two Nusselt number relationships (eqns. 3.11 and 3.12) for hot wire probes having typical dimensions is that these equations do not include the effect of conduction to the support needles. The next section will present methods of calculating the magnitude of these losses.

Not all sensors are cylindrical in shape. Hot film sensors, although sometimes configured as a low-aspect-ratio circular cylinder, often have quite different shapes. It is common to use the King's law expression of eqn. 3.7 for film probes because it gives good results, but other equations are sometimes used for hot film probes in water. Giovanangeli (1980), for example, used a modified form of the Collis and Williams law, and Bonis and van Thinh (1973) used Kramer's law for water measurements.

The flat surface hot film probe is most often used to measure wall shear stress, and a heat transfer law for it was first proposed by Ludweig (1950). It was later modified by Owen and Bellhouse (1970) and Geremia (1972) and expressed by Geremia as

$$E_b^2 = A_3 + B_3 \tau_w^{1/3}$$

where $\tau_w$ is the wall shear stress, and constants $A_3$ and $B_3$ are determined experimentally for the type of flat surface probe used.

### Heat transfer in rarefied gases

When a hot wire probe is used in ambient temperature room air, the temperature of the fluid at the surface of the sensor is equal to the temperature of the sensor. But a rarefied gas has an intermolecular spacing that is large enough to interfere with heat conduction between the sensor and the fluid, and a discontinuity between the two temperatures exists. This temperature jump influences the heat transfer in rarefied gas flows and causes a calibration curve taken in a continuum to be useless in a rarefied gas without a correction being applied.

For any material we can express the mean free path length, $\lambda$, which is the average distance a molecule travels between collisions by assuming a Maxwellian distribution of molecular velocities, as

$$\lambda = \frac{0.707}{4\pi r^2 n}$$

where $r$ is the radius of the gas molecules, and $n$ is the molecular density (number of molecules per unit volume).

A dimensionless number, the Knudsen number, Kn, can be used to characterize rarefied gas flows. It is defined as

$$\text{Kn} = \frac{\lambda}{x}$$

where $x$ is the characteristic length of the sensor. For a rarefied gas, Kn $\approx$ 1.

For any gas the relationship between the Knudsen number, Mach number, and Reynolds number is

$$\text{Kn} = \left(\frac{\pi\gamma}{2}\right)^{0.5} \frac{M}{\text{Re}}$$

where $M$ is the Mach number, and $\gamma$ is the specific heat ratio.

The Nusselt number can be corrected for rarefied gas effects by using the following relationship proposed by Collis and Williams (1959):

$$\frac{1}{\text{Nu}_c} = \frac{1}{\text{Nu}_m} - 2\text{Kn}$$

where $Nu_m$ is the measured Nusselt number for the probe and $Nu_c$ is the Nusselt number corrected for rarefied gas effects. This equation allows a probe to be calibrated in continuum flow and then used in rarefied gas flow.

### The lower limit for velocity measurements

Typical hot wire probe calibration curves show a marked decrease in sensitivity at low velocities. This leads us to ask what the lower limit is for practical hot wire and hot film measurements.

If a hot wire or hot film sensor is placed in a quiescent fluid, convection currents (see Brodowicz and Kierkus [1966] for an interesting photograph of this effect) are formed because of the buoyancy of the heated fluid. Thus, the sensor loses heat by convection even in still air. This free convection flow tends to mask any low-velocity forced convection cooling, resulting in a flattening of the lower end of the calibration curve.

As the fluid velocity is reduced, the probe becomes less sensitive to velocity until measurements are no longer possible. A rule of thumb for air based on practical experience in using a wide variety of equipment places the lower-velocity limit for conventional hot wire anemometers at about 0.15 m/s or 0.20 m/s (Jorgensen, 1979). In water the lower-velocity limit is several orders of magnitude less.

In the mixed convection regime where neither free nor forced convection are negligible, Collis and Williams (1959) tested large-aspect-ratio sensors and found the lowest Reynolds number for which they could be used to be

$$Re_{min} = 1.85 Gr^{0.39} \left(\frac{T_m}{T_0}\right)^{0.76} \tag{3.13}$$

where $T_m$ is the mean film temperature, $T_0$ is the fluid ambient temperature, and Gr is the Grashof number, a dimensionless expression relating buoyant to viscous forces and defined as

$$Gr = \frac{g\beta(T_s - T_0)x^3}{\nu^2}$$

where $x$ is the characteristic length, $\nu$ is the kinematic viscosity, $g$ is the acceleration of gravity, and $\beta$ is the volume coefficient of expansion, defined as

$$\beta = -\frac{1}{\rho}\left(\frac{\partial\rho}{\partial T}\right)_p$$

Equation 3.13 is valid for Re < 0.1.

At low velocities the calculation of the Reynolds number is complicated by the existence of the component of velocity due to free convection that is superimposed on the forced convection component. If free convection is neglected, sizable errors can result. Hatton, James, and Swire (1970) chose to use this concept of error due to a free convection component to define the lower-velocity limit of the anemometer. They chose a lower-velocity limit for which the free convection component would cause more than a 10% error in the calculation of the Reynolds number based on the forced convection velocity only. These lower limits are

Parallel flow:    $Re = 10Ra^{0.418}$
Cross flow:      $Re = 2.2Ra^{0.418}$
Contra flow:     $Re = 9Ra^{0.418}$

and they may be used in the range $10^{-2} < Re < 40$ and $10^{-3} < Ra < 10$. The contra flow case is not valid when $0.25 < Ra^{0.418}/Re < 2.5$.

In the above equation the Rayleigh number, Ra, is defined as

$$Ra = Gr \cdot Pr$$

In parallel flow the free convection flow is parallel to and in the same direction as the forced convection flow. In cross flow the two flows are at right angles to one another, and in contra flow free and forced convection oppose one another.

At least one manufacturer offers a hot wire anemometer for low-velocity measurements having a vibrating sensor oscillating in the plane of the forced convection vector. Electronic circuitry calculates the velocity with the probe moving both toward and against the forced convection vector and electronically separates the fluid velocity from the sensor velocity. The lower-velocity limit is said to be extended by this instrument, which is suitable for mean velocity measurements only.

#### Humidity effects

The effect of humidity is usually ignored in laboratory measurements because relative humidity varies little inside a room over a period of days or weeks. The effect of humidity on hot wire anemometry measurements was investigated experimentally by Schubauer (1935), using a wind tunnel in which relative humidity could be varied without changing the temperature. In these tests the relative humidity was varied from 25% to 70% with the temperature maintained at 25°C. Relative humidity changes were found to cause velocity errors of up to 6%.

### 3.4 Conduction effects in sensors

The effects of lateral heat conduction within a sensor and to support needles, insulation coverings, and substrates is often assumed to be negligible, but this is usually far from true.

#### Conduction to support needles

An end-loss correction method proposed by Davies and Fisher (1964) can be used to correct for heat loss from the sensor to the support needles. Using the Fourier law of heat conduction, we obtain the expression for heat loss to both support needles:

$$\text{Conduction heat transfer rate to support needles} = 2k_s A \left|\frac{dT_s}{dx}\right|_{l/2}$$

where $k_s$ is the thermal conductivity of the sensor material, and $|dT_s/dx|_{l/2}$ is the absolute value of the slope of the temperature profile at the sensor ends. The temperature profile for a cylindrical sensor was given by eqn. 3.6, with the constants $K_1$ and $K_2$ given by eqns. 3.4 and 3.5, respectively. Equation 3.6 is differentiated to find the slope of the sensor temperature profile at the sensor end where $x = l/2$ and substituted into the conduction heat loss equation. This gives

Conduction heat transfer rate to support needles

$$= 2k_s A \frac{K_2}{\sqrt{|K_1|}} \tanh\sqrt{|K_1|} \frac{l}{2} \quad (3.14)$$

If the constants $K_1$ and $K_2$ can be evaluated, then eqn. 3.14 can be used to find the fraction of heat leaving the sensor by conduction to the support needles.

The constant $K_2$ can be calculated, but $K_1$ contains $h$, which cannot easily be measured. Instead, $K_1$ is expressed in terms of the Nusselt number for a sensor of infinite length (eqn. 3.11):

$$\text{Nu}_1 = \frac{I^2 R_s}{\pi l k_s (T_s - T_f)}$$

and the Nusselt number of the moderate aspect ratio sensor being used (eqn. 3.8)

$$\text{Nu}_2 = \frac{hd}{k_f}$$

The constant $K_1$ becomes, with eqn. 2.5,

$$K_1 = \frac{4k_f}{d^2 k_s} \text{Nu}_1 \left(\frac{R_s - R_f}{R_s} - \frac{\text{Nu}_2}{\text{Nu}_1}\right)$$

But $K_1$ still cannot be evaluated because $\text{Nu}_2$ contains $h$. Another way to express $\text{Nu}_2$ is to observe that the numerator of $\text{Nu}_1$ is the sensor heat generation rate. If the heat transfer rate to the support needles is subtracted from this numerator, a new expression for $\text{Nu}_2$ results. Then the Nusselt number ratio becomes

$$\frac{\text{Nu}_2}{\text{Nu}_1} = 1 - \frac{2k_s A}{I^2 R_s} \frac{K_2}{\sqrt{|K_1|}} \tanh\left(\sqrt{|K_1|} \frac{l}{2}\right)$$

These last two equations express $K_1$ in terms of the Nusselt number ratio, but they cannot be used to find $K_1$ explicitly, however, and an iterative method must be used instead. Once $K_1$ is found, the numerical values for it and $K_2$ can be substituted into eqn. 3.14, and the total conduction heat loss to the support needles can be calculated.

The accuracy of this end-loss correction technique depends upon an accurate measurement of $l$, $R_s$, $R_0$, $d$, and $k_s$, none of which can be determined easily. For example, it will be difficult to measure the sensor length, $l$, because even with the aid of a microscope the exact sensor attachment point is difficult to find. The measurement of resistances $R_s$ and $R_0$ are complicated by the resistance of the support needles, welded joints, and leads, which cannot easily be separated from the sensor resistance. Also, the sensor diameter often is not constant, and $k_s$ for a wire may be different from that of a polished specimen of the same material (Schmidt and Cresci, 1971).

The heat loss by conduction to the support needles can be found experimentally by placing the probe in an evacuated chamber. The convective heat loss is negligible in a vacuum, and, because radiation heat loss is also negligible, the heat generated electrically is balanced by the heat lost to the support needles (Rajasooria and Brundin 1971, and Schmidt and Cresci 1971). Then the heat transfer rate to the support needles is

$$\text{Heat transfer rate to support needles} = I^2 R_s$$

**Conduction along the sensor**

An uneven velocity profile along the length of the sensor will cause heat to move by conduction from the hotter parts of the sensor to the cooler, resulting in a temperature profile that is more uniform than would otherwise be expected.

A time constant for the conduction time along the sensor has been expressed by Corrsin (1963) as

$$\tau = \frac{l^2}{\alpha}$$

where $\alpha$ is the thermal diffusivity of the sensor, defined as

$$\alpha = \frac{k_s}{\rho c}$$

where $\rho$ is the density of the sensor material, $c$ is its specific heat, and $k_s$ is its thermal conductivity.

**Conduction through insulation coatings**

Insulation coverings are used over sensors to prevent damage by abrasion and to insulate them electrically from the fluid. In addition, contaminants on a sensor can act as a nonuniform, time-dependent insulation coating.

If a cylindrical sensor is covered with a coating of uniform thickness, the heat transfer rate, $q$, through the coating is

$$q = \frac{2\pi kl(T_1 - T_2)}{\ln(r_2/r_1)} \tag{3.15}$$

where subscripts 1 and 2 refer to the surface of the sensor and the outer surface of the coating, respectively, $k$ is the thermal conductivity of the coating, and $r$ is the radius.

If the sensor is long, the end conduction losses can be assumed to be negligible; then the heat generation rate is equal to the heat rate lost by conduction through the coating or

$$q = I^2 R_s \tag{3.16}$$

Combining eqns. 3.15 and 3.16 to solve for the temperature of the outer surface of the coating, we obtain

$$T_2 = T_1 - \frac{I^2 R_s}{2\pi kl} \ln \left(\frac{r_2}{r_1}\right)$$

The heat transfer rate through the insulation coating of a flat-surface probe is

$$q = \frac{kA}{\Delta x} (T_1 - T_2) \tag{3.17}$$

where $\Delta x$ is the thickness of the insulation covering, and $A$ is the surface area of the film. If, as before, all heat generated by the sensor is assumed to pass by conduction through the insulation coating, eqns. 3.16 and 3.17 can be combined:

$$T_2 = T_1 - \frac{I^2 R_s \Delta x}{kA}$$

The slight difference in temperature between the sensor and the surface of the insulating coating can be accounted for in the calibration.

### Heat conduction into substrates

An important heat loss path in noncylindrical hot film probes is conduction through the substrate adjacent to the sensor, because not only can temporal variations in substrate temperature influence the sensor calibration, but conduction into the substrate causes errors in the measurement of low-frequency velocity fluctuations.

Suppose a hot film probe is taken from a cool cabinet and placed in a warm room. An hour may pass before the temperature of the quartz substrate reaches the room temperature, but if the sensor is calibrated soon after leaving the cabinet, the calibration curve will be in error after the substrate reaches room temperature. Hutton and Gammon (1976) found the substrate heating effect caused by holding the probe body of a flat-surface film probe between the thumb and forefinger required about 10 minutes to disappear, and calibration of the probe immediately after being held by the fingers gave errors.

Substrate heating can also effect the frequency response of the anemometer. A fluctuating fluid velocity cools both sensor and substrate, and although the effect is felt immediately by the sensor through the air, it is experienced later through the substrate. Since only low-frequency heat waves penetrate any distance into the glass substrate, the hot film probe having a thick substrate will respond accurately to moderate and high-frequency velocity fluctuations and have impaired low-frequency response. A hot wire probe, on the other hand, has excellent low-frequency response.

### 3.5 Radiation heat transfer from sensors

The radiation heat transfer rate from the sensor to the surroundings is usually assumed to be negligible, and for most applications the radiation heat loss is much smaller than losses by convection to the fluid or conduction to the support needles. But to be certain, the magnitude of the radiation heat transfer should be calculated.

In order to calculate the radiation heat transfer rate for a cylinder of length $l$, the expression for the heat transfer rate away from a differential element by radiation (eqn. 3.2) is modified to obtain

$$\text{Radiation heat transfer rate} = \sigma \pi d l \epsilon (T_s^4 - T_{\text{sur}}^4) \qquad (3.18)$$

where $\sigma$ is the Stefan–Boltzmann constant having a value of $\sigma = 5.67 \times 10^{-8} \text{ W/m}^2 {}^\circ\text{K}$, $T_{\text{sur}}$ is the temperature of the surroundings, and $\epsilon$ is the emissivity of the sensor material. This equation can be used if the emissivity is known.

The emissivity of the sensor material can be found in tables of material properties, or it can be measured by using a method developed by Schmidt and Cresci (1971). In this method a long piece of sensor material is placed inside an evacuated bell jar and heated electrically. Since the aspect ratio is large, conduction losses to the wire supports can be neglected. Convection losses can also be neglected, and this means the heat generated electrically is balanced by the heat lost by radiation to the surroundings, which are at ambient temperature. Thus

$$\epsilon = \frac{I^2 R_s}{\sigma \pi d l (T_s^4 - T_{\text{sur}}^4)}$$

Substitute the calculated values of effective emissivity into eqn. 3.18 to find the radiation heat transfer for a hot wire probe. Schmidt and Cresci found these tests to give values of emissivity two and one-half times higher than handbook values and attributed this discrepancy to the unpolished and somewhat dirty surface of the wire.

## 3.6 Effect of ambient temperature changes

The convective heat transfer rate from a sensor depends upon the difference in temperature between the sensor and the fluid; although the constant temperature anemometer automatically maintains a constant sensor temperature with great accuracy, fluid temperature is not always constant. For example, the temperature of the air in a closed-circuit wind tunnel will often increase several degrees during operation, causing errors that may invalidate the data taken.

### Temperature and velocity sensitivity

Analysis of temperature sensitivity begins with the Collis and Williams heat transfer law (eqn. 3.9), developed for large-aspect-ratio wires in air and having all fluid properties evaluated at the mean film temperature. This is combined with the definition of Nusselt number (eqn. 3.11) and Ohm's law to give

$$\frac{E_s^2}{\pi l k R_s (T_s - T_f)} \left(\frac{T_m}{T_f}\right)^{-0.17} = A_1 + B_1 \left(\frac{\rho U d}{\mu}\right)^{n_1} \tag{3.19}$$

where $\mu$ is the absolute viscosity. The fluid properties such as $k$, $\mu$, and $\rho$ vary with the fluid temperature and can be expressed as the product of a reference property and a temperature ratio raised to a power. The first two are given by Collis and Williams (1959) as

$$k = k_o \left(\frac{T_m}{T_o}\right)^{0.80}$$

$$\mu = \mu_o \left(\frac{T_m}{T_o}\right)^{0.76}$$

where the subscript $o$ denotes an arbitrary reference condition. The equation of state for a perfect gas at constant pressure is also used:

$$\rho = \rho_o \left(\frac{T_m}{T_o}\right)^{-1}$$

In these equations the reference property is defined as the value of the property at the reference temperature. Substituting into eqn. 3.19, we get

$$\frac{E_s^2 T_f^{0.17}}{(T_s + T_f)^{0.97}(T_s - T_f)} = C_1 + D_1 \left[\frac{U}{(T_s + T_f)^{1.76}}\right]^{n_1} \tag{3.20}$$

where

$$C_1 = \frac{\pi l k_o R_s}{2^{1.14} T_o^{0.80}} A_1$$

$$D_1 = \frac{\pi l k_o R_s}{2^{1.14} T_o^{0.80}} \left[\frac{\rho_o d}{\mu_o} (2T_o)^{1.76}\right]^{n_1} B_1$$

where for constant temperature operation both $T_s$ and $R_s$ are constant.

Each variable in this expression is then assumed to be composed of the sum of a large mean part and a small fluctuating part or

$$E_s = \overline{E}_s + e_s, \qquad T_f = \overline{T}_f + t_f, \qquad U = \overline{U} + u$$

where $1 \gg e_s/\overline{E}_s$, $1 \gg t_f/\overline{T}_f$, and $1 \gg u/\overline{U}$. The overbar indicates mean quantities. Substitute these into eqn. 3.20 and neglect terms containing the product of two or more fluctuating quantities. In addition, the equation can be modified to obtain factors having the form of a small quantity added or subtracted from unity. Then approximations of these can be formed with binomial series expansions. A typical example is

$$\left(1 + \frac{t_f}{T_s + \overline{T}_f}\right)^{-0.97} \approx 1 - 0.97 \frac{t_f}{T_s + \overline{T}_f}$$

Our equation then contains both mean and fluctuating parts with higher-order terms neglected. Finally, eliminate $C_1$ from this equation by substituting eqn. 3.20, expressed in terms of mean values only. The result can be written in a form similar to eqn. 1.1:

$$e_s = S_{\text{vel}}u + S_{\text{temp}}t_f \tag{3.21}$$

where the velocity sensitivity, $S_{\text{vel}}$, is expressed as

$$S_{\text{vel}} = \left\{ \frac{n_1 \pi l k_o R_s}{2^{1.97}\overline{E}_s \overline{U} T_o^{0.80}} \left[ \frac{\rho_o \overline{U} d}{\mu_o} (2T_o)^{1.76} \right]^{n_1} B_1 \right\}$$
$$\cdot \frac{(T_s + \overline{T}_f)^{0.97 - 1.76n_1}(T_s - \overline{T}_f)}{\overline{T}_f^{0.17}}$$

and the temperature sensitivity, $S_{\text{temp}}$, is

$$S_{\text{temp}} = -\frac{1}{2}\left\{ \frac{1.76 n_1 \pi l k_o R_s}{2^{0.97}\overline{E}_s T_o^{0.80}} \left[ \frac{\rho_o \overline{U} d}{\mu_o} (2T_o)^{1.76} \right]^{n_1} \right.$$
$$\left. \cdot B_1 \frac{T_s - \overline{T}_f}{\overline{T}_f^{0.17}(T_s + \overline{T}_f)^{0.03 + 1.76n_1}} \cdot \overline{E}_s \left( \frac{0.17}{\overline{T}_f} + \frac{1}{T_s - \overline{T}_f} - \frac{0.97}{T_s + \overline{T}_f} \right) \right\}$$

The signs associated with the temperature sensitivities indicate that an inadvertent decrease in fluid temperature will be interpreted by the unsuspecting user as an increase in velocity.

If the heat balance equation for mean values is substituted into these sensitivity equations to eliminate $\overline{E}$, a decrease in sensor temperature will be seen to cause a decrease in the velocity sensitivity and an increase in the temperature sensitivity. This allows the hot wire anemometer to be used as a temperature sensor, an application to be discussed in a subsequent section.

### Temperature compensation

There are several temperature compensation methods in use: They include maintaining constant overheat ratio, maintaining constant temper-

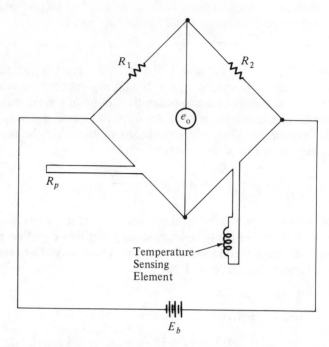

Figure 3.6. A Wheatstone bridge anemometer with a coil used as a temperature compensation device. The coil takes the place of the variable resistor in the Wheatstone bridge. Because temperature variations that effect the sensor also alter the resistance of the coil, the bridge balance is not disturbed. The coil is placed in the flow field near the hot wire sensor, using the type of probe shown in Figure 3.7.

ature difference, and applying correction factors to the anemometer output signal, either graphically or by using a suitable equation. Although none give perfect compensation, they allow measurements that might otherwise be impossible.

The most common temperature compensation technique requires manual adjustment of sensor temperature to maintain constant overheat ratio as the ambient temperature varies. Fluid temperature can be measured by using the hot wire anemometer as a resistance thermometer, or a thermometer or thermocouple can be used. This compensation method is useful when the temperature varies slowly or when time or expense limitations preclude the use of other methods.

The constant overheat ratio technique need not be a manual operation, however, and a popular method is to use a coil of wire as an overheat resistor in the Wheatstone bridge circuit (Figure 3.6). The coil is placed in the fluid near the probe, but not close enough to cause probe heat to influence the coil. If the diameter and length of the wire used in the coil are chosen properly, the coil will have the correct resistance yet undergo virtually no electrical heating. Then a change in ambient temperature will have the same

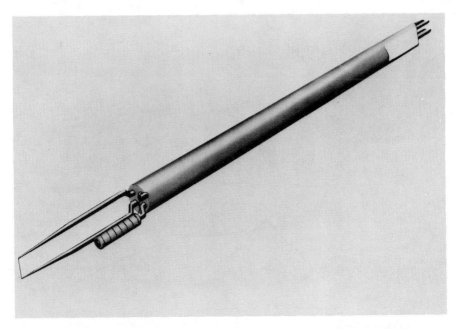

Figure 3.7. A temperature compensated hot wire probe with a coil of wire used as the compensating element. Reprinted with permission from Dantec Elektronik.

effect on both sensor and coil, and the Wheatstone bridge will remain in balance. A probe designed for this purpose is shown in Figure 3.7.

Alternatively, a second standard probe can be used as a temperature sensor along with a resistor network (Drubka, Tan-atichat, and Nagib, 1977) as shown in Figure 3.8. Again, the second probe and resistor network take the place of the overheat resistor in the Wheatstone bridge, and the resistance of the network is such that the sensor of the second probe is barely heated. Since a sensor operated at a low overheat ratio has little sensitivity to velocity, this second probe will not be influenced by fluid velocity changes. Drubka, Tan-atichat, and Nagib reported maximum errors of only 1% in velocity by using this method when the ambient temperature was varied by 20°C.

Temperature changes can also be compensated by maintaining a constant temperature difference between ambient and sensor temperatures. This method was investigated by Drubka, Tan-atichat, and Nagib (1977), but found to give less satisfactory results than the constant overheat ratio method.

Another approach is to apply a correction factor to the output signal of the anemometer to correct for ambient temperature changes. A correction factor proposed by Bearman (1971) is

$$E_{b_c} \approx E_b \left[ 1 - \frac{T_{o_1} - T_{o_2}}{2(T_s - T_{o_1})} \right]$$

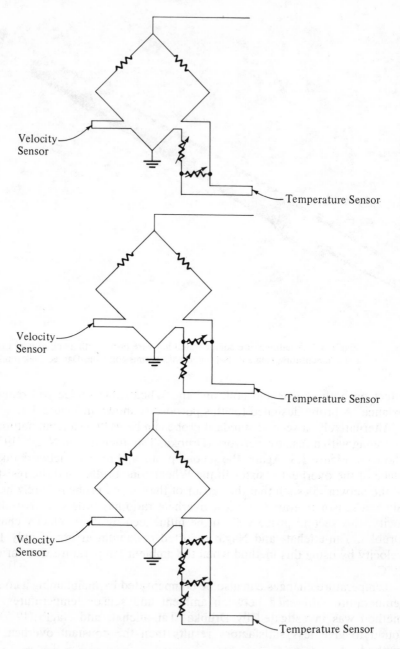

Fig. 3.8. Several circuits for temperature compensation of velocity measurements using a second hot wire probe and a resistance network. Reprinted with permission from R. E. Drubka, J. Tan-atichat, and H. M. Nagib, Analysis of temperature compensating circuits for hot wires and hot films, *DISA Info.*, 22 (1977), 5–14.

where $E_b$ is the anemometer bridge voltage recorded after the ambient temperature changed, and $E_{b_c}$ is the corrected bridge voltage. The subscripts 1 and 2 refer to conditions before and after the previous and new ambient temperature change, respectively, and the subscript $o$ refers to an arbitrary reference condition. Another correction factor, proposed by Kanevce and Oka (1973), is

$$E_{b_c}^2 \approx E_b^2 \left( \frac{R_s - R_{o2}}{R_s - R_{o1}} \right)$$

A final temperature correction method requires the construction of a series of calibration curves over the range of the ambient temperature variations expected. If ambient temperature is monitored separately, each output voltage reading can be converted to velocity by interpolating the calibration curves.

## 3.7 Calibration of probes

There is enough variation in the physical characteristics of seemingly identical probes to require a separate calibration curve of output voltage versus velocity for each. In these sections we explain how to calibrate probes in a variety of fluids, in which the probe is either held stationary or oscillated back and forth, and how to construct calibration curves from the data.

### The steady-state calibration method

Of the two ways to calibrate a probe, the most popular is the "steady-state" method, in which the probe is held stationary while the fluid passes over it, or the probe is traversed at constant speed through a quiescent fluid. In either case the anemometer output voltage is recorded at discrete velocities, or the velocity is varied slowly with time while the output voltage is recorded continuously.

The steady-state calibration method is simple to perform, and the needed equipment is commercially available or can be handmade. A disadvantage is the need for a long integration time to process the data for turbulent flow. Comte-Bellot (1977) found that to achieve an accuracy of 1% an integration time of about 10 seconds was required for each data point when calibration was performed in turbulent flow at 20 m/s.

There are two ways to perform the steady-state calibration: either pass the fluid over a stationary probe, or traverse the probe through the still fluid. Moving the probe through the fluid assures a constant velocity profile at the sensor, but this is achieved at the expense of limited probe travel.

For gases it is usually more convenient to move the fluid past the probe by using a wind tunnel. It is usually easier to place the probe at different angles to the velocity vector if an open jet wind tunnel is used.

For liquids the fluid can be moved past the probe by immersing a small pump in a tank of liquid and directing a jet at known velocity over the probe.

Figure 3.9. A stationary tank with overhead rotating arm for calibration of hot film probes in water. The rotating arm can be driven at a variety of speeds, and the location of the probe on the arm can be changed as well. Notice the plastic sheeting used for the tank walls. Reprinted with permission from J. Anhalt, Device for in-water calibration of hot-wire and hot-film probes, *DISA Info.*, 15 (1973), 25–26.

Alternatively, the probe can be placed outside of a liquid-filled tank adjacent to a hole in the side or bottom to allow the stationary probe to be immersed in the outgoing jet. The velocity can be calculated from a measurement of the height of the liquid in the tank at any instant. The temperature of the water usually remains constant in this calibration technique, but if the water cools with time, the test can be run quickly. Rubatto (1970) allowed a tank to empty in 10–15 minutes, and the water temperature did not vary by more than ±0.1°C. For constant-head operation build an overflow pipe into the tank. An automatic temperature control device may be needed to maintain constant temperature. Pichon (1970) used a thermostatically controlled heater or cooling coils, depending upon which was necessary, to maintain constant water temperature.

Another advantage of the hole-in-tank calibration method is that the turbulence intensity produced in the flow is low. After the fluid in the tank is allowed to stand, the only source of turbulence is that generated by the edges of the hole, and this can be minimized with proper design. A disadvantage is the difficulty of measuring the water height when only a small amount of water remains in the tank. Consequently, 1 m/s is the practical lower limit for this method (Pichon, 1970).

Another method is to house the liquid in an open circular container that rotates under a stationary probe. The velocity can be varied by changing the location of the probe above the tank. Continuous measurements are not possible because the probe causes the liquid to rotate. For accurate calibration the tank should be brought up to speed, the probe dipped in, a measurement taken, and the probe removed – all in less than one revolution of the tank, if possible.

Equally common is calibration by traversing a probe through a stationary liquid. One method is to use a liquid container in the shape of a long, narrow, open channel acting as a towing tank with rails upon which a small trolley carrying the probe travels. The probe extends a constant depth below the liquid surface, and the trolley can be powered by a geared motor or a weight attached to the trolley by a cord passing over a small pulley. Typical towing tanks are described by Pichon (1970). A second method is to attach the probe to an arm that rotates horizontally over an open tank of the liquid. Continuous measurement of velocity is still not possible because the liquid begins to move with the probe after one revolution (Bertrand and Couders, 1978). This method is illustrated in Figure 3.9, where the walls of the tank are seen to be made of plastic sheeting. A dome (not shown) covers the entire tank to prevent the entry of dust. The depth of the probe below the liquid surface as well as the probe angle and speed can be varied from outside the dome without interrupting the test. For the configuration shown the probe speed can be varied from 5 cm/s to 600 cm/s. Mercury slip rings are used to transfer power and control signals to the rotating arm (Anhalt, 1973).

### Construction of steady-state calibration curves

A calibration curve of output voltage versus velocity is the simplest display of the calibration data, but a linearized calibration curve can be constructed by first assuming the data to follow King's law (eqn. 3.7). A straight line will be obtained if $E_b^2$ versus $U^n$ is graphed. This curve is somewhat ambiguous, because the slope of the linearized curve depends upon the exponent chosen for King's law. This exponent is a function of velocity and varies over the range $0.40 < n < 0.50$. Perry and Morrison (1971a) plotted identical calibration data (see Figure 3.10), using exponents of 0.40, 0.45, and 0.50 to illustrate the effect this has on the slope of the calibration curve.

Mulhearn and Finnigan (1978) used King's law in the form

$$E_b^2 = A + BU^n$$

to make a least squares fit of their data from a typical commercial X-array probe. They calculated the constants $A$ and $B$ for values of the exponent in the range $0.2 < n < 2.0$ and found that for minimum standard deviation the exponent for one sensor of the probe was $n = 0.76$, and the exponent for the other sensor was $n = 0.72$.

### Calibration by shaking the probe

A calibration technique primarily used to find the slope of the calibration curve is performed by oscillating the probe in the streamwise di-

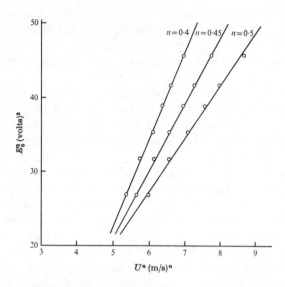

Figure 3.10. Curves of identical calibration data, assuming different values for the exponent in King's equation. Reprinted with permission from A. E. Perry and G. L. Morrison, Static and dynamic calibration of constant-temperature hot-wire systems, *J. Fluid Mech.*, 47 (1971), 765–777.

rection in a moving fluid. It is less common to oscillate the probe in a stationary fluid or to oscillate the fluid in the vicinity of a stationary probe.

This method is not new (Schubauer and Klebanoff, 1946) and has been developed by Perry and Morrison (1971a), who measured both the rms value of probe velocity and the bridge voltage, the anemometer being ac coupled to the rms voltmeter at the time. The ratio of these two rms values is the velocity sensitivity.

Because the output voltage is a nonlinear function of velocity, the probe must be shaken over a velocity range small enough to allow bridge voltage variations to be considered to be a linear function of velocity. For this reason Perry and Morrison (1971a) limited the probe velocity to less than 10% of the free stream velocity.

The sensor should not move in relation to the support needles during shaking. Perry and Morrison limited shaking frequency to less than 15 Hz to avoid this. Relative motion between the sensor and the support needles can be detected by repeated calibration by using different overheat ratios, and discrepancies noted in the data. Because bowing of the sensor is more exaggerated at higher sensor temperatures, one would expect a greater relative motion to occur then. Another technique is to use different shaker frequencies at the same fluid velocity; both methods were used by Perry and Morrison.

Unfortunately shaking mechanisms that allow adjustable displacement and speed are not available commercially. A typical shaker mechanism, designed

Figure 3.11. A probe-shaking mechanism designed to move the probe in an elliptical path. Reprinted with permission from P. J. Mulhearn and J. J. Finnigan, A simple device for dynamic testing of x-configuration hot-wire anemometer probes, *J. Phys. E.: Sci. Instr.*, 11 (1978), 679–681.

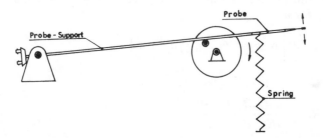

Figure 3.12. A probe-shaking mechanism designed to move the probe in a semicircular arc. Reprinted with permission from A. A. Gunkel, R. P. Patel, and M. E. Weber, A shielded hot-wire probe for highly turbulent flows and rapidly reversing flows, *Ind. Engr. Chem. Fund.*, 10 (1971), 627–681.

by Mulhearn and Finnigan (1978), is shown in Figure 3.11. A less complicated design is shown in Figure 3.12.

The shaking calibration method can also be used to find the steady-state calibration curve by plotting the peak bridge voltage versus the instantaneous velocity of the fluid passing the probe. This method gives identical results to those obtained by the steady-state calibration method (Bruun, 1976). Sreenivasan and Ramachandran (1961) found that for cylinders having an aspect ratio of 17, a stationary cylinder had the same heat transfer rate as one vibrating with velocities as high as 20% of the mean velocity.

One seldom calibrates a probe by shaking it in a quiescent fluid, but this technique can be used if the quantities of the fluid are limited.

### 3.8 The hot wire sensor frequency response

When a hot wire sensor experiences a sudden change in velocity, a finite time is required for the sensor temperature to reach its new value. This is caused by a number of effects, the most important being heat storage in the sensor itself.

In order to develop an expression for the frequency response of a heated wire sensor, the heat balance equation for a heated sensor, given by eqn. 3.3, is used in an analysis after Hinze (1959, pp. 76–83). If the sensor is assumed to be long enough for the temperature profile to be constant, and heat loss by radiation is assumed to be negligible, this equation becomes

$$\frac{I^2 \rho_r}{A} = \rho c A \frac{dT_s}{dt} + \pi dh (T_s - T_f) \tag{3.22}$$

Using eqns. 2.3 and 2.5, we get

$$I^2 R_s = \frac{\rho c A l}{\alpha R_o} \frac{dR_s}{dt} + \frac{\pi dh l}{\alpha R_o} (R_s - R_f) \tag{3.23}$$

Then eqn. 3.8, the Nusselt number, is used with Kramer's law (eqn. 3.10) and solved for $h$.

$$h = \frac{k_f}{d} \left[ 0.42 \mathrm{Pr}^{0.20} + 0.57 \mathrm{Pr}^{0.33} \left( \frac{Ud}{\nu} \right)^{0.5} \right]$$

Substituting into eqn. 3.23, we get

$$I^2 R_s = \frac{\rho c A l}{\alpha R_o} \frac{dR_s}{dt} + \frac{\pi l k_f}{\alpha R_o}$$

$$\cdot \left[ 0.42 \mathrm{Pr}^{0.20} + 0.57 \mathrm{Pr}^{0.33} \left( \frac{Ud}{\nu} \right)^{0.5} \right] (R_s - R_f) \tag{3.24}$$

Next the usual assumption regarding turbulent flow is made; that is, all fluctuating quantities can be expressed as the sum of a mean component, represented as an uppercase letter with an overbar, and a fluctuating component, represented by a lowercase letter. Then velocity and sensor resistance become $U = \overline{U} + u$ and $R_s = \overline{R}_s + r_s$, respectively. Because the sensor current is assumed to be constant, $I_s = \overline{I}_s$. Neglecting higher-order terms, subtracting eqn. 3.24 for mean values from eqn. 3.24 containing both mean and fluctuating values, and using binomial series expansions to give, for example,

$$\sqrt{\overline{U} + u} \approx \sqrt{\overline{U}} + \frac{u\sqrt{\overline{U}}}{2\overline{U}}$$

give us the following first-order linear ordinary differential equation:

$$\frac{dr_s}{dt} + \frac{\alpha R_o}{\rho cAl}\left\{\frac{\pi lk_f}{\alpha R_o}\left[0.42Pr^{0.20} + 0.57Pr^{0.33}\left(\frac{\overline{U}d}{\nu}\right)^{0.5}\right] - \overline{I}^2\right\}r_s$$

$$= \left(\frac{\pi k_f}{2\rho cA\overline{U}}\right)0.57Pr^{0.33}\left(\frac{\overline{U}d}{\nu}\right)^{0.5}(\overline{R}_s - R_f)u$$

The time constant for this equation is

$$\tau_{cca} = \frac{\rho cAl(\overline{R}_s - R_f)}{\overline{I}^2\alpha R_f R_o} \tag{3.25}$$

Calculation of the time constant for a typical hot wire sensor will show it to be on the order of a few hundred hertz – far too low to allow the probe to be used for turbulence measurements in the constant current mode without frequency compensation.

# 4 ELECTRONIC CIRCUITRY

The user of hot wire anemometry is often an expert in fluid mechanics but may know relatively little about electronics; and although a detailed knowledge of the electronics package is not required for accurate measurements, an understanding of its operation develops confidence in the measurements taken. The two main parts of the electronics package are a Wheatstone bridge circuit and feedback amplifier. In this chapter we discuss each.

## 4.1 The Wheatstone bridge

A fundamental part of the electronics package for both constant current and constant temperature anemometers is the Wheatstone bridge circuit. In fact, it can be used alone, without feedback electronics or signal processing, for some applications.

### The basic Wheatstone bridge circuit

The Wheatstone bridge is used in many electronic instruments to give an error signal proportional to the difference between a variable signal and a reference signal. Its use in hot wire anemometry is a typical application.

In its simplest configuration the Wheatstone bridge circuit is composed of four resistors, as shown in Figure 4.1. The equivalent resistance, $R_{eq}$, of the Wheatstone bridge is the resistance seen by the voltage source:

$$R_{eq} = \frac{(R_1 + R_4)(R_2 + R_3)}{R_1 + R_2 + R_3 + R_4} \tag{4.1}$$

The power consumed by the bridge is

$$\text{Power} = \frac{E_b^2}{R_{eq}}$$

The total current passing through the bridge circuit is

$$I_b = \frac{E_b}{R_{eq}}$$

An important relationship in hot wire anemometry is the current through one resistor – for example, $R_4$. It is

$$I_4 = \frac{E_b}{R_1 + R_4} \tag{4.2}$$

84

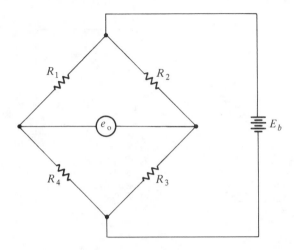

Figure 4.1. A typical Wheatstone bridge circuit.

A Wheatstone bridge is said to be balanced if its resistors have values such that the unbalanced voltage $e_o$ is zero. Analysis of the circuit shows this to be the case if

$$A = \frac{R_2}{R_1} = \frac{R_3}{R_4}$$

where $A$ is the bridge ratio.

The equivalent resistance for a balanced bridge is

$$R_{eq} = \frac{A}{1 + A} (R_1 + R_4) \tag{4.3}$$

The power consumption for a balanced bridge is

$$\text{Power} = \frac{E_b^2(1 + A)}{A(R_1 + R_4)} \tag{4.4}$$

### The Wheatstone bridge in hot wire anemometry

The Wheatstone bridge configuration used in hot wire anemometry is shown in Figure 4.2. It is identical to the bridge configuration of Figure 4.1, except that resistor $R_4$ is replaced by the probe resistance $R_p$, and resistor $R_3$ is replaced by an adjustable resistor $R_{adj}$.

The probe resistance is a composite of the resistance of the sensor plus the resistance of the support needles, probe body, leads, and probe cable. Variable contact resistance can lead to problems. For example, if the contact resistance at the connectors in the probe and the cable vary while velocity measurements are taken, the data will be useless. The contact resistance at

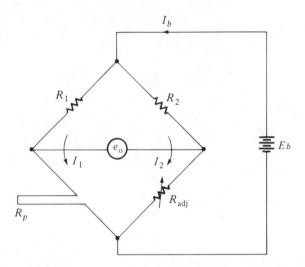

Figure 4.2. A Wheatstone bridge with one resistor replaced by a hot wire probe and used as a hot wire anemometer. The adjustable resistor is used to set the overheat ratio.

the variable resistor may also vary. To reduce this problem, high-quality switches and connectors having low and stable contact resistances should be used. In fact, it is more common for an electronics package malfunction to be caused by the mechanical failure of a switch or potentiometer than the failure of an electronic component.

Although capacitance in the probe arm of the Wheatstone bridge has negligible effect on performance (Perry and Morrison, 1971), probe cable inductance, $L_c$, has a significant effect. If the inductance of one arm of the bridge is larger than another, a bridge balanced for slow variations in probe resistance or bridge current will become unbalanced as the frequency of the fluctuations increases. Inductance in the probe arm of the bridge is caused by the inductance of the probe cable. Cable inductance can be reduced by using a shorter cable. Anemometers having the bridge circuit mounted inside the probe body to minimize cable length have been offered commercially, and Wehrmann (1968) mounted not only the bridge circuit but also the feedback amplifier in the probe body.

A second way to reduce the effect of inductance is to compensate it by adding inductance to the bridge arm containing the variable resistor. An inexpensive and simple method is to attach the adjustable resistor to the Wheatstone bridge by a coaxial cable of the same length and type as the probe cable, since equal lengths of the same type of cable can be assumed to have the same inductance. Alternatively, a variable inductor can be placed in series with the adjustable resistor in the bridge circuit, as shown in Figure 4.3, and adjusted to optimize frequency response.

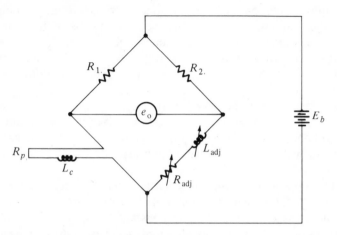

Figure 4.3. A Wheatstone bridge anemometer with variable inductor for compensation of the probe cable inductance.

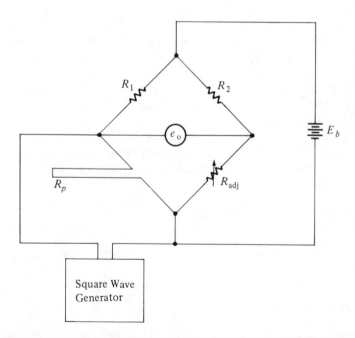

Figure 4.4. A square wave generator can be used to test the frequency response of a Wheatstone bridge.

The frequency response of the Wheatstone bridge circuit can be measured by using the "square wave test." Replace the probe with a noninductive resistor of the same resistance and superimpose an electric current in the form of a square wave on the probe current by placing a square wave generator in parallel with the probe, as shown in Figure 4.4. With the bridge in operation, observe the output voltage to see how faithfully it follows the square wave signal. To measure the frequency response characteristics of the bridge circuit with an operational probe, place it in a fluid at the same velocity to be measured and repeat the square wave test. Calculation of the time constant will be discussed in a later section.

### The Wheatstone bridge as a constant current anemometer

The Wheatstone bridge circuit of Figure 4.2 can be used as a constant current anemometer by maintaining constant bridge current. This can be done with a constant current power supply (Sajben, 1965), or a battery can be placed in series with a large resistor so that variations in equivalent resistance of the bridge will be small compared to the total resistance seen by the current source. This second method was used by Dryden and Kuethe (1929) and Spangenberg (1955). These approaches are illustrated in Figure 4.5.

For constant current operation place the probe in a flow of known velocity and adjust the probe heating current for a suitable unbalance voltage. Vary the fluid velocity over the range of interest while recording the calibration data. Then without changing the bridge current put the probe in the flow to be measured.

A high heating current is needed in high-speed flow for adequate velocity sensitivity, but if the velocity is inadvertently reduced, the sensor temperature will rise and may cause a burnout.

For constant current operation several of the bridge equations developed earlier take a different form. The equivalent resistance of a bridge for constant current operation can be found by rewriting eqn. 4.1 and using the notation of Figure 4.2 to give

$$R_{eq} = \frac{(R_1 + R_p)(R_2 + R_{adj})}{R_1 + R_2 + R_p + R_{adj}}$$

Other bridge characteristics can be expressed by writing the bridge equations in terms of the bridge current, $I_b$.

For constant current operation the power consumption is

$$\text{Power} = I_b^2 R_{eq}$$

It can be shown that power consumption is reduced if $R_2$ and $R_{adj}$ are much larger than the probe resistance.

From eqn. 4.2 the probe current, $I_p$, is

$$I_p = \frac{I_b(R_2 + R_{adj})}{R_1 + R_2 + R_p + R_{adj}}$$

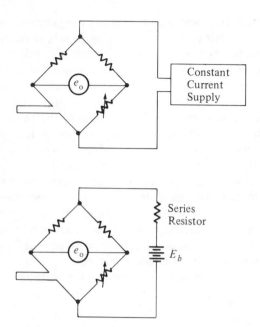

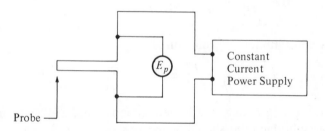

Figure 4.5. A constant current Wheatstone bridge circuit can be powered either by a constant current power supply or a large resistance in series with a battery.

Figure 4.6. It is possible to use a probe as a constant current anemometer without a Wheatstone bridge, as this circuit shows. The output voltage is the voltage drop across the probe.

The Wheatstone bridge circuits of Figure 4.5 can be used as constant current anemometers without modification. The frequency response will be limited by the frequency response of the probe, and, although this is no more than a few hundred hertz, this instrument can be used as a low-cost, easily fabricated anemometer.

A Wheatstone bridge circuit is not necessary for constant current operation. Instead the probe can be powered directly by a constant current supply. The output voltage is obtained by measuring the voltage drop across the probe, as shown in Figure 4.6. This circuit can be used for steady-state measurements, but it cannot be compensated for probe cable reactance, making it useless for high-frequency measurements.

### The Wheatstone bridge as a constant temperature anemometer

The Wheatstone bridge circuit of Figure 4.2 can be used as a constant temperature anemometer by manually varying the bridge voltage, $E_b$, to maintain bridge balance. For constant temperature operation set the adjustable resistor to a higher value than would be correct for a balanced bridge based on the resistance of the unheated probe. Apply the bridge voltage and increase it until the sensor heats. This increases the sensor resistance and brings the bridge into balance. Remove any bridge unbalance caused by velocity variations by readjusting the bridge voltage. The resulting bridge voltage is directly proportional to the fluid velocity.

Whereas the modern constant temperature anemometer uses a feedback amplifier to automatically maintain the sensor temperature constant, the manual method can be used in measurements of the heat transfer from long, thin wires (Champagne, Sleicher, and Wehrmann, 1967) or for classroom demonstrations of the constant temperature anemometry principle when more sophisticated equipment is not available.

The bridge ratio for constant temperature operation is

$$A = \frac{R_2}{R_1} = \frac{R_{\text{adj}}}{R_p}$$

Using this equation and eqn. 4.3, we obtain the equivalent resistance

$$R_{\text{eq}} = \frac{A}{1 + A}(R_1 + R_p)$$

The power consumption is, from eqn. 4.4,

$$\text{Power} = \left(\frac{1 + A}{A}\right)\frac{E_b^2}{R_1 + R_p}$$

Notice that for long battery life in field experiments, a Wheatstone bridge having a large bridge ratio should be used.

In the design of bridge circuits for constant temperature anemometers, it is common for the adjustable resistor to have high resistance compared to that of the probe in order to minimize the effect of variable contact resistance in the rotary switches of the adjustable resistor. This is accomplished by using a high bridge ratio.

If extensive lengths of probe cable are needed, the cable inductance must be compensated; one method is to mount a fixed resistor in place of the adjustable resistor at the end of a cable that is identical to the one used by the probe.

## 4.2 The constant current anemometer

The constant current anemometer was the first type used for fluid velocity measurements and the first to be sold commercially. Although the modern

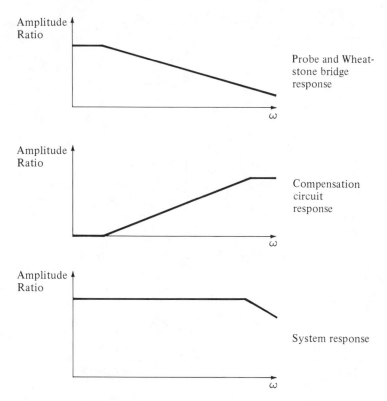

Figure 4.7. The principle of operation of the frequency compensator of a constant current anemometer. The compensation circuit gives greater amplification at higher frequency to compensate for the opposite tendency of the probe. In this graph, $\omega$ is the circular frequency.

constant current anemometer has a frequency response that is high, it has been replaced almost exclusively by the constant temperature anemometer, except for its continued use as a resistance thermometer.

The modern constant current anemometer uses a Wheatstone bridge with a compensation circuit added to improve system frequency response. it is illustrated in Figure 1.4. Without compensation, the frequency response of the constant current anemometer is low; at higher frequencies the amplitude ratio (ratio of the amplitude of the output voltage change to the velocity fluctuation amplitude) decreases. The compensation circuit maintains constant amplitude ratio by amplifying the bridge output voltage, so that amplification is greatest at higher frequencies. Because the slope of the bridge amplitude ratio versus frequency is maintained almost constant by the compensator, it acts as a differentiator with an output signal proportional to the slope of the input signal. Of course, for proper operation the characteristics of the compensator must be matched to those of the bridge. The effect of the compensator on amplitude ratio is shown in Figure 4.7.

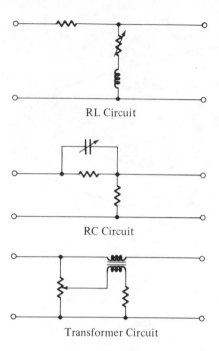

RL Circuit

RC Circuit

Transformer Circuit

Figure 4.8. Three types of passive networks that can be used to increase the frequency response of a constant current anemometer (Kovasznay, 1954).

A passive frequency-compensation technique for a constant current ane-mometer is one of the networks shown in Figure 4.8. Three variations are the *RC, RL*, and transformer circuits (Kovasznay, 1954), but all attenuate the bridge output signal. To overcome this, these circuits are usually fol-lowed by an amplifier. Integrated circuits can be used in place of the passive networks to differentiate the output voltage from the Wheatstone bridge for frequency compensation.

The compensation technique for constant current operation is complicated by several effects. The time constant for a hot wire sensor having constant heating current is given by eqn. 3.25, and for a purely resistive bridge circuit this is also the frequency response of the constant current anemometer with-out frequency compensation. But eqn. 3.25 shows that the time constant depends upon the sensor temperature. Because the sensor temperature for a constant current anemometer varies with velocity, the frequency response of the system is seen to be velocity dependent, and the compensation circuit gives optimum compensation at only one velocity. Thus, the compensation circuit must be readjusted for each new velocity range encountered and will not be effective in turbulent flow. According to Comte-Bellot (1976), these errors become significant if the turbulence level exceeds about 5%.

A minor advantage of the constant current anemometer is the simple adjustment of sensor temperature by varying the bridge current at the power supply. In contrast, the sensor temperature of a constant temperature anemometer is adjusted by turning off the sensor heating current, changing the setting of the adjustable resistor, and turning on the probe heating current again.

## 4.3 The constant temperature anemometer

The constant temperature anemometer is the primary research tool of the experimentalist in fluid mechanics, although its position is being challenged by the laser Doppler anemometer. The constant temperature anemometer has a frequency response that is high, and this instrument is readily available on the commercial market at a relatively low price. The analyses presented in this chapter are adapted from Perry and Morrison (1971).

### ✳ System operation

The feedback amplifier in a constant temperature anemometer senses the unbalance voltage of the Wheatstone bridge and adds current to the bridge to restore balance. This amplifier is usually composed of several individual amplifiers in series. Typically, a differential amplifier amplifies the unbalance voltage from the bridge, and the current level of this signal is increased with a current booster. This basic system is illustrated in Figure 4.9. Two of the differential amplifier input controls are connected to the bridge, and the bridge unbalance signal is amplified by a factor of 500, or sometimes more. A front-panel amplifier gain control is sometimes provided to optimize the system frequency response.

The third input terminal is used as an offset control. This allows the output terminals to show a voltage difference when there is no voltage difference at the input terminals when a suitable offset voltage is applied. This offset voltage is needed to allow the circuit to begin to operate when the anemometer is turned on. The offset control is either fixed at the factory or adjustable from the front panel.

A differential amplifier is designed to have high common-mode rejection, which means that a voltage difference between the two input terminals will be amplified, whereas a voltage common to both terminals will not. Thus, a change in the overall bridge voltage will not be sensed by the amplifier, but it will still respond to bridge off-balance voltage changes.

In order to insure adequate heating current to maintain constant temperature in high-velocity applications, the current from the differential amplifier is increased by a unity-gain amplifier acting as a current booster. Schematic diagrams of simple hot wire anemometer circuits are given by Freymuth (1967), Wyngaard and Lumley (1967), Weidman and Browand (1975), Hembling (1980), and Ming Ho (1982).

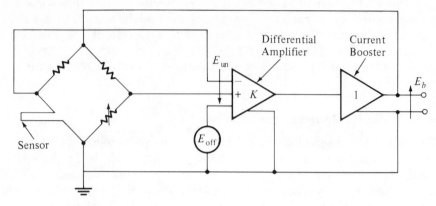

Figure 4.9. The circuit diagram of a constant temperature anemometer. Reprinted with permission from A. E. Perry and G. L. Morrison, A study of the constant-temperature hot-wire anemometer, *J. Fluid. Mech.*, 47 (1971), 577–599.

### The frequency response of the constant temperature anemometer

The major advantage of the constant temperature anemometer is that its frequency response is high when compared with the frequency response of an uncompensated constant current anemometer. The following derivation of the frequency response for a constant temperature anemometer has been adapted from Hinze (1959, pp. 116–119).

The differential equation for a heated sensor was given by eqn. 3.24. For constant temperature operation the variables are written as the sum of mean and fluctuating parts. Thus, $U = \overline{U} + u$, $I_s = \overline{I}_s + i_s$, and $R_s = \overline{R}_s + r_s$. Eliminating terms containing $i_s^2$, $i_s r_s$, and $u r_s$, because they are small compared to others, and making the additional approximation of eqn. 3.22 gives

$$2 i_s \overline{I}_s \overline{R}_s + \overline{I}_s^2 r_s = \frac{\pi l k_f}{\alpha R_o} \left[ 0.42 \mathrm{Pr}^{0.20} r_s + 0.57 \mathrm{Pr}^{0.33} \sqrt{\frac{\overline{U} d}{\nu}} \right.$$

$$\left. \cdot \left( r_s + \frac{\overline{R}_s - R_f}{2 \overline{U}} u \right) \right] - \frac{\rho c A l}{\alpha R_o} \frac{d r_s}{dt} \quad (4.5)$$

Next, a relationship expressing the current vs. resistance characteristics for the feedback amplifier is

$$i_s = - g \overline{I}_s r_s$$

where $g$ is the amplifier transconductance. Differentiating this equation and substituting it into eqn. 4.5 gives

$$\frac{d i_s}{dt} + \frac{\alpha R_o \overline{I}_s^2 [2 g \overline{R}_s (\overline{R}_s - R_f) + R_f^2]}{\rho c A l (\overline{R}_s - R_f)} i_s$$

$$= 0.57 \mathrm{Pr}^{0.33} \sqrt{\frac{\overline{U} d}{\nu}} \frac{\pi l k_f (\overline{R}_s - R_f)}{2 \alpha R_o \overline{U}} u$$

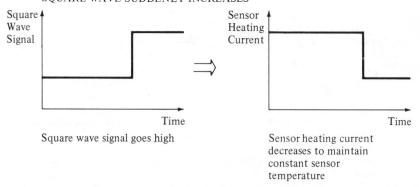

Figure 4.10. A comparison of the constant temperature anemometer response to a change in fluid velocity and to the square wave test. The anemometer responds in the same way to a sudden decrease in velocity as it does to a sudden increase in the square wave current.

The time constant for this differential equation is

$$\tau_{cta} = \frac{\rho c A l (\overline{R}_s - R_f)}{\alpha R_o \overline{I}_s^2 [2gR_s(\overline{R}_s - R_f) + R_f^2]}$$

This is the expression for the time constant for a constant temperature anemometer.

### The square wave test

The square wave test allows measurement and optimization of the system frequency response. For this test place a square wave generator in parallel with the probe, as shown in Figure 4.4, and add a small-amplitude square wave to the sensor heating current. For greater realism operate the sensor in steady flow to allow the square wave signal to be superimposed on a large mean signal.

The square wave test is based on the assumption that heating and cooling

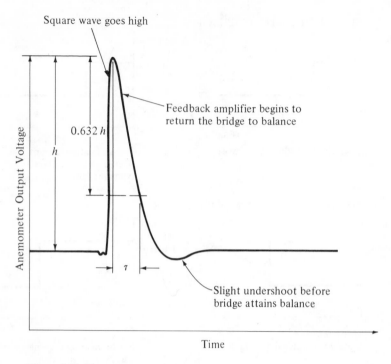

Figure 4.11. The constant temperature anemometer output signal during the square wave test, showing how the time constant, $\tau$, for the system is measured.

of the sensor by varying the fluid velocity is thermodynamically identical to heating and cooling of the sensor by varying the heating current. But because convective cooling is a surface phenomenon, whereas electrical heating takes place deep within the sensor material, these two phenomena are only approximations of one another.

When the square wave current is introduced, the sensor current will go high and then drop to its original value as the feedback amplifier reduces the heating current to balance the bridge. A sudden decrease in fluid velocity has the same effect, as shown in Figure 4.10. Alternatively, when the square wave current goes low, the feedback amplifier will increase the heating current to bring the sensor resistance back to its original value.

The time constant for the system can be calculated by measuring the time required for the output voltage to drop from its peak amplitude to an amplitude that is 63.2% less, as shown in Figure 4.11.

### Other frequency response tests

Although the square wave test is universally used to measure frequency response, two alternative schemes have been used. One method uses a type of surface heating that more closely approximates convective cooling.

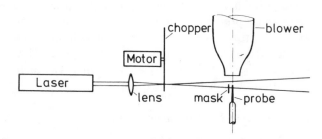

Figure 4.12. The measurement of frequency response by using a laser to heat the sensor. A motor-driven chopper disk is used to modulate the laser beam.
Reprinted with permission from H. Fiedler, On data acquisition in heated turbulent flows, in *Proc. Dyn. Flow Conf.*, pp. 81–100, Skovlunde, Denmark, 1978.

Typical of these techniques is heating of the sensor by acoustics, microwaves, or lasers. A second method uses hydrodynamic variations in shock tubes or behind cylinders from which vortices are shed.

An early frequency response test using surface heating was the use of microwaves in the 10-GHz range by Kidron (1966). Acoustical heating of the sensor surface was analyzed theoretically by Lueck and Osborn (1979) and shown to be valid in fluids having high-enough values of thermal diffusivity to cause rapid heating of the fluid surrounding the sensor. This implies that acoustical heating should not be used to measure the frequency response of a sensor in water where thermal diffusivity is small.

A technique having great promise uses laser heating of the sensor to determine its frequency response. Comte-Bellot (1975) used a 1-W argon laser with its beam deflected from side to side with a Bragg cell to vary the light intensity at the sensor. A motor-driven chopper disk was used by Fiedler (1978) to interrupt the laser beam cyclically. A mask was used to allow heating either the sensor alone or the sensor and the support needles together, as shown in Figure 4.12.

Ideally, the frequency response of a sensor should be measured by using fluid velocity variations. One technique is to mount the sensor in a shock tube to respond to the wave front of a shock wave as it passes (Lu, 1979). The step change in output voltage can be compared to the step change in velocity to calculate the sensor frequency response. Davis (1970) found this shock tube method to give results identical to the square wave technique.

The frequency response of a hot wire anemometer can also be measured by placing the sensor in a turbulent flow of known intensity. This test is repeated for flows of different intensity, and the frequency response calculated. In this method, developed by Perry and Morrison (1971a), a vortex street is generated by using a circular cylinder. A probe located downstream from the cylinder is traversed laterally across the wake. This test is repeated for cylinders of different diameters. In this technique the turbulence intensity is a function of the Reynolds number Re as well as the nondimensional distances, $x/d$ and $y/d$, where $x$ and $y$ are coordinate directions and $d$ is the diameter of the cylinder producing the vortices. If Re and $x/d$ are held constant, the probe can be traversed in the $y$-direction behind the cylinder, and turbulence intensity versus lateral distance can be graphed. This is done for cylinders of different diameter, and any separation between the resulting curves is an indication of poor frequency response.

### 4.4 The alternating-current hot wire anemometer

An early development in hot wire anemometry was the use of an alternating current to heat the sensor. Two variations are possible. One is a "constant current" anemometer having the sensor heated by a constant rms alternating current; the other, a constant temperature type with the rms value of heating current proportional to bridge unbalance. Advantages of the alternating-current anemometer have been nullified by subsequent developments in hot wire anemometry. For example, when ac anemometers were developed, the insulated hot film probe had not been invented, and measurements in water were made with uninsulated hot wire probes. Probe fouling by bubbles was a major problem. Alternating-current operation of the sensor reduced bubble formation somewhat and also overcame instabilities due to polarization of the water (Patterson, 1958). Also, before the invention of transistors, feedback amplifier design was difficult for early builders of hot wire anemometers, whereas the amplifier for an ac constant temperature anemometer was a commercially available audio amplifier and two transformers (Shepard, 1955).

An example of the constant current type of ac hot wire anemometer was developed by Stevens, Borden, and Strausser (1956) for measurements with an uninsulated hot wire probe in water. A signal generator, power amplifier, and transformer were used to power a Wheatstone bridge with a sinusoidal voltage. Bridge unbalance was sensed with a transformer and amplifier, and the output signal was displayed on an oscilloscope.

An early ac constant temperature anemometer was designed by Shepard (1955); it had self-excited oscillation of the loop composed of the Wheatstone bridge, amplifier, and transformers. A conventional Wheatstone bridge circuit was used along with a 10-W audio amplifier with a gain of about 40. It was isolated by input and output transformers and designed to feed back a signal proportional to the bridge unbalance voltage. This regenerative feedback design had an oscillation frequency of about 8000 Hz, high enough so that the sensor heat-generation rate remained constant. Later, Grant, Stewart, and Moilliet (1962) developed an alternating-current hot film anemometer for measurements with uninsulated hot film probes in water.

# 5 FLUIDS

Although air is the most common fluid in which hot wire measurements are taken, other liquids and gases are often used. In this chapter we look at the special techniques required to make measurements in different fluids. Techniques for air measurements are considered first, but only the more specialized air applications are covered; less commonly used fluids are given a more thorough treatment.

## 5.1 Measurements in air

Because instruction manuals for the hot wire anemometer often assume air to be the fluid under investigation, only its use for atmospheric measurements and calibration techniques for air are discussed here.

### Atmospheric measurements

When making atmospheric measurements, the equipment may be left unattended for days or weeks while measurements are taken. This coupled with the logistics of moving the equipment and personnel from the laboratory to the field test site and back again pose serious problems.

Probe replacement at the test site can be a problem if the probes must be mounted on a tower. If the tower is not designed with a pivot point near the ground for rotation to a horizontal position to allow probe replacement from the ground, the tower must be climbed to retrieve the probe and again to replace it after repair or recalibration – sometimes under adverse weather conditions. For shipboard measurements the tower may sway during the climb. For safety a body harness should be worn while climbing, although it must be repeatedly attached and detached as one moves.

Electrical power may not be available at the field test site if located outside the city. Batteries, motor-driven portable generators, or solar cells must then be used. If batteries are chosen, all equipment must be capable of battery operation and have low current drain to extend battery life.

Shipboard use of hot wire probes for atmospheric measurements poses special problems because of salt encrustation of the sensor. Electron microscope photographs of salt-encrusted hot wire sensors are shown in Figure 5.1. Tests by Schacher and Fairall (1976) indicate that, although large coatings of salt are quickly formed on the sensor, negligible change in calibration occurred. Instead, the increased diameter and mass of the wires when coated

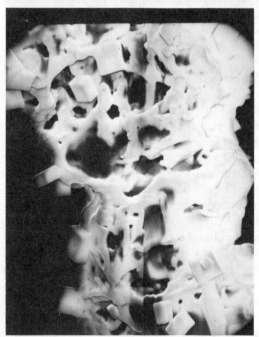

Figure 5.1. Scanning electron microscope photographs of laboratory produced salt layers on 4.5-μm-diameter tungsten wires. The diameter of the salt layer is 42 μm. Magnification is 70X for the upper photograph and 1400X in the lower. Reprinted with permission from C. W. Fairall and G. Schacher, Frequency response of hot wires used for atmospheric turbulence measurements in the marine environment, *Rev. Sci. Instr.*, 48 (1977), 12–17.

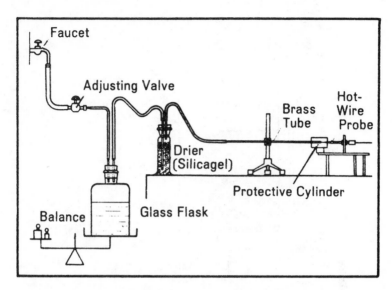

Figure 5.2. A bottle can be filled with a constant flow of water to produce a low-velocity air flow for hot wire probe calibration. Reprinted with permission from P. Almquist and E. Legath, The hot-wire anemometer at low air velocities, *DISA Info.*, 2 (1965), 3–4.

caused breakage due to wind and vibration. The hot wire probe lifetime in shipboard applications was found to be one and one-half days for velocity measurements and three days for temperature measurements, and it was necessary to clean the sensors at the end of each day to achieve these results. An additional problem in shipboard measurements was the continued rocking of the vessel, which caused masking of velocity measurements below about 5 Hz.

For atmospheric measurements on Mars with the Viking Mars Lander, Henry and Greene (1974) chose hot film probes with sensors covered with Teflon or aluminum oxide to reduce abrasion by blowing dust. Constant current operation at low overheat ratio was chosen to conserve battery power.

### Calibration in air

Small recirculating and nonrecirculating wind tunnels specifically designed for probe calibration are commercially available, and they allow quick and sometimes automatic calibration and data recording over a wide velocity range. Simple calibration wind tunnels can also be handmade at little cost. Usually the velocity standard against which the probe is calibrated is a pitot tube.

For calibration at low velocity, a handmade calibration facility may be required. A common method is to force air from a large glass bottle by introducing water at a constant rate, as shown in Figure 5.2. The water flow

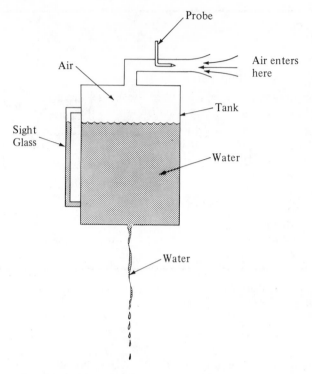

Figure 5.3. The draining of a container of water can be used to produce a low-velocity flow of air suitable for hot wire probe calibration.

rate is measured by periodically recording the weight of water in the bottle, and a small container of silica gel can be used to dry the air. This method was used by Almquist and Legath (1965) over a velocity range from 0.15 m/s to about 2.0 m/s.

A similar technique is to allow water to flow from the bottom of a large enclosed tank with a probe inserted in an opening at the top through which air enters, as shown in Figure 5.3. An advantage of this method is that dry air passes over the probe. Of course, the rate at which water leaves the container is not constant but directly proportional to the height of the water in the tank.

Hot wire probes can also be calibrated at low velocity by using neutrally buoyant soap bubbles carried along with the flow. Use trial and error to find a mixture of nitrogen and hydrogen that gives neutrally buoyant soap bubbles. Squeeze them by hand from a flexible plastic bottle through a small tube previously dipped in soap solution. Make bubbles that are about 2 cm in diameter, and let them be carried along in the low-velocity air past vertical lines marked on a background material. A stroboscope can be used to illuminate the bubbles, and they can be photographed with a camera with the shutter held open. This technique was used by Ling and Lowe (1981) to measure velocities on the order of 0.3 m/s with an accuracy estimated at ± 5%.

## 5.2 Measurements in water

Water is the most commonly used liquid for velocity measurements with the hot wire anemometer and offers valuable training for measurements in other liquids. In this section we look at techniques for making freshwater measurements, followed by special considerations for saltwater use. Finally, we discuss the many techniques for calibration of probes in water.

### Freshwater measurements

There are several precautions necessary when water measurements are taken. First, the water should be connected electrically to the anemometer ground. Notice that one terminal of the probe is connected to the anemometer ground in the schematic diagram of Figure 4.9. If the water is also grounded to the instrument, there will be less likelihood that a large potential difference will exist between the probe and the water and less chance of an electrical breakdown of the insulation coating covering the sensor.

The connectors of the probe cable, its support, and the probe should have waterproof seals. If water enters, the anemometer output voltage will vary erratically, necessitating the removal of probe and cables for drying. Some manufacturers use waterproof connectors for the cable and the probe, but they can also be waterproofed by coating them with the type of plastic sealant ordinarily used to seal bathtubs.

### Saltwater measurements

Because salt water usually occurs in an ocean or estuarine environment, the problems of taking measurements in salt water are often compounded by the difficulty of working at a field test site. An interesting example of shipboard use of hot film probes is shown in Figure 5.4. Three teflon-coated hot film probes are mounted to the deck of a research submersible, *Ben Franklin*. The tubing framework above the probes protects against breakage.

The metal cases that usually house the instruments usually do not protect them from damage by corrosion when exposed to humid, salty air. If a hot wire anemometer must be used on an open deck, it can be encased in clear plastic sheeting having all openings sealed with tape or heat. If the sheeting is loose, the controls can be operated through the plastic.

The most serious problem of hot film probe measurements in a marine environment is that the quartz protective layer covering the sensor can be dissolved by salt water. The lifetime of quartz-insulated hot film probes used in seawater was found by Forman (1971) to be about 20 hours. This was

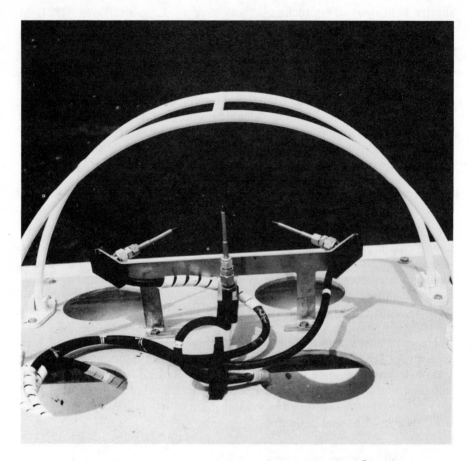

Figure 5.4. For some undersea applications it may be possible to mount the probes directly to the deck of the research vehicle. Here, three hot film sensors are mounted under a protective framework on the deck of the research submersible *Ben Franklin*. Reprinted with permission from K. M. Forman, Design and integration of turbulence experiments for mesocaphe 'Ben Franklin', ASME paper, 69-WA/UnT-9.

extended by sputtering a coat of teflon onto the sensor. Forman showed that probes insulated this way survive in the ocean for as long as 950 hours at depths of up to 600 m, plus an additional 256 hours of surface towing. When these teflon-coated sensors finally failed, microscopic examination showed that seawater had entered through small pores in the teflon, causing a weakening of the bond between the teflon and the quartz. This caused the teflon to lift away from the sensor.

### The calibration of probes in water

Liquid calibration methods were discussed in Section 3.7, but some specialized techniques have been developed for water.

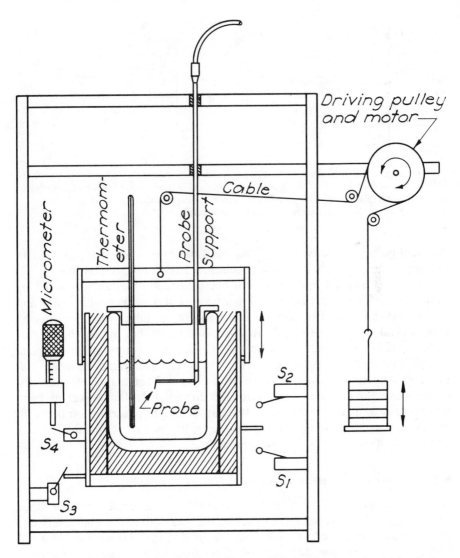

Figure 5.5. In this low-speed water-calibration device a container of water is raised slowly under a stationary hot film probe. Reprinted with permission from R. P. Dring and B. Gebhart, Hot-wire anemometer calibration for measurements at very low velocity, *J. Heat Trans.*, 91 (1969), 241–244.

A low-speed calibration method using a stationary probe was developed by Dring and Gebhart (1969) and is illustrated in Figure 5.5. The calibration liquid is placed in an open bucket partially submerged in a larger open tank of water containing heaters or cooling coils to maintain constant temperature. The bucket is suspended by a cable passing over a pulley with a counterweight attached to the other end of the cable. The pulley is driven by an electric motor at low rpm to slowly raise the bucket. A hot film probe is

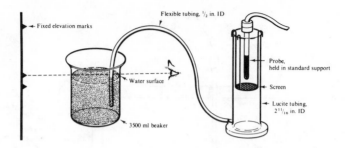

Figure 5.6. A beaker held at eye level allows a constant flow of water to pass by a stationary probe in this unusual low-speed calibration device for hot film probes in water. Reprinted with permission from B. Gallagher, Calibration of probes in water at velocities under one meter per second, *DISA Info.*, 14 (1973), 19–20.

mounted pointing downward, so that the water moves past the probe. With this apparatus, velocities from 0.13 mm/s to 5.08 mm/s are possible.

An unconventional low-velocity water calibration method proposed by Gallagher (1973) is illustrated in Figure 5.6. Fill a beaker with water and insert a siphon tube. Connect the other end of the tube to the bottom of a glass cylinder containing a downward-pointing probe. When the beaker water level is above the bottom of the cylinder, the water will be siphoned into the cylinder, slowly filling it, and allowing a calibration to be made. For constant velocity the water level in the beaker must remain a constant distance above the cylinder water level. To do this, hold the beaker by hand and move it to maintain the water level at the height of a mark on the wall, as seen by sighting along the top of the water by eye. Constant velocity occurs when the water overflows the cylinder.

## 5.3 Measurements in polymer solutions

When small quantities of long-chain polymers are added to water or other solvents, the shear stress decreases, and the thermodynamic characteristics of the liquid are altered as well. This means that the performance of a hot film probe in a polymer solution will differ from its performance in water.

### Probes for use in polymer solutions

If velocity measurements in a polymer solution are needed, one usually chooses either a cylindrical, wedge, or conical film probe (Figure 5.7). Each type, however, has its limitations when used in polymer solutions. For example, the conical film probe was reported by Fabula (1968) to be superior to the wedge probe, which had a calibration curve of abnormal shape, due presumably to the macromolecular buildup on the leading edge of the wedge. Vortex shedding can occur from a cylindrical probe at Reynolds numbers greater than about 50, and probes found to be insensitive to

Figure 5.7. The conical hot film probe offers superior resistance to fouling because of its shape. A thin, metal film is deposited in a ring around the tip of a conical quartz rod. Reprinted with permission from Dantec Elektronik.

velocity below that Reynolds number become sensitive after the onset of vortex shedding (Friehe and Schwarz, 1969). Friehe and Schwarz also found that the anemometer output signal is not the typical sinusoidal shape usually observed when vortex shedding occurred behind cylindrical probes in water. Instead, the output variations were irregular, as if the probe were experiencing turbulent flow.

If hot film probes are not available, hot wire probes can be used. Astarita and Nicodemo (1969) were able to measure in polymer solutions with an uninsulated wire probe without bubble-formation problems.

### Heat transfer from probes in polymer solutions

When making velocity measurements, an adequate change in heat transfer rate with velocity is desired for good sensitivity. Friehe and Schwarz (1969) found the Nusselt number to remain constant for some concentrations of polymer solution over certain velocity ranges, indicating that the probe is insensitive to velocity in those regimes. In addition, Smith, Merrill, Mickley, and Virk (1967) found the heat transfer coefficient to be unstable, varying at random by a factor of as much as three, at velocities in the range from 600 cm/s to 900 cm/s. In addition, Metzner and Astarita (1967) found that probes became increasingly insensitive to velocity at high speed or in highly elastic fluids.

In general, long-chain polymer molecules interfere with the molecular interaction necessary for good heat transfer between the probe and the fluid (Kalashnikov and Kudin, 1973), causing a decrease in the heat transfer coefficient. Long-chain molecules gradually deteriorate with time, causing the heat transfer coefficient to increase. This is shown in Figure 5.8.

James and Acosta (1970) found heat loss from probes at low speeds in polymer solutions to be the same as in water, but at higher speeds differences were noted. This critical speed was found to vary with concentration, as shown in Table 5.1.

When a probe is placed in a polymer solution at an angle to the flow, the heat transfer rate will increase; in most other fluids the opposite is true. This effect, reported by Friehe and Schwarz (1969) and shown graphically in Figure 5.9, also exhibits angular ambiguity that allows one velocity reading to represent several angular values.

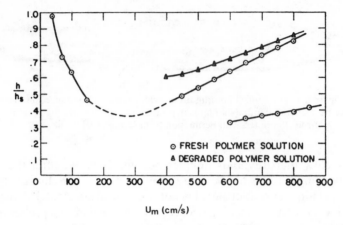

Figure 5.8. The variation of the polymer solution heat transfer coefficient with velocity and degradation time. In this graph, $U_m$ is the mean velocity, $h$ is the heat transfer coefficient of the polymer solution, and $h_s$ is the heat transfer coefficient for the solvent. Reprinted with permission from K. A. Smith, E. W. Merrill, H. S. Mickley, and P. S. Virk, Anomalous pitot tube and hot film measurements in dilute polymer solutions, *Chem. Engr. Sci.*, 22 (1967), 619–626.

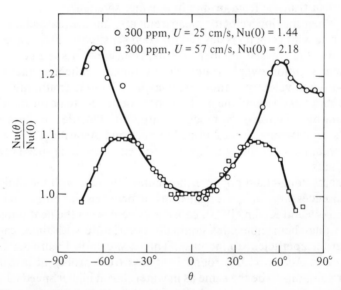

Figure 5.9. The variation in Nusselt number with sensor angle for a polymer solution. In this graph $Nu(\theta)$ is the Nusselt number of the angled sensor, $Nu(0)$ is the Nusselt number of the sensor when normal to the mean velocity vector, and $\theta$ is the yaw angle. Reprinted with permission from C. A. Friehe and W. H. Schwarz, Deviations from the cosine law for yawed cylindrical anemometer sensors, *J. Appl. Mech.*, 90 (1968), 655–662.

Table 5.1. *The variation in critical
Reynolds number, $Re_c$, with polymer
concentration*

| Polymer concentration | $Re_c$ |
| --- | --- |
| 7.4 | 6 |
| 15.7 | 4.5 |
| 30.2 | 3 |

*Source:* James and Acosta (1970).


## 5.4 Measurements in mercury

Many of the problems encountered in mercury measurements are not present
with other liquids. These include the danger of mercury poisoning, special
probe-fouling conditions, and errors caused by magnetic fields.

### Probes for use in mercury

Mercury is an excellent conductor of electricity and requires the use
of insulated probes. In addition, one must be sure that no breakdown of the
insulation occurs. Sajben (1965) insulated his wire sensors by coating them
with a 25-$\mu$m layer of enamel, and although this gave adequate insulation,
these probes were highly sensitive to fouling.

Probe fouling in mercury has always been a formidable problem. Sajben
(1965) was never able to eliminate fouling and instead devised a method that
allowed the use of fouled probes. Fouling was reduced to an acceptable level
by Hoff (1968), who used hot film probes having a thin, metal film deposited
over the quartz insulation covering the sensor. This allowed wetting of the
probe surface by the mercury and reduced the formation of deposits. Of the
metals used for deposition, Robinson and Larsson (1973) chose vanadium,
and Hoff used both copper and gold. Hoff found a copper film to erode after
only 2 hours of use, whereas a gold film lasted from 5 to 100 hours. Hoff
also found the metal film to seal tiny holes in the quartz insulation – holes
that otherwise seemed responsible for random sensor-resistance variations
characteristic of standard hot film probes. Platnieks (1971) used 9-$\mu$m diameter tungsten wires 2 mm long and vacuum deposited with silica as an
insulation. The silica was sprayed from a distance of about 10 cm while the
sensor was twice rotated 120°.

Because mercury is denser than many other liquids, extreme hydrodynamic forces are possible, and this dictates the use of rugged, adequately
supported probes. Wedge or conical probes are an excellent choice. Cylin-
drical-sensor hot film probes having an OD of 0.152 mm were used by Hoff
(1968) and found to break if a velocity of 0.76 m/s was exceeded. They were,

however, able to withstand a velocity of 1.52 m/s after being strengthened by applying epoxy at the sensor ends. Wedge and parabolic hot film probes were used by Hoff at mercury velocities as high as 4.33 m/s without breakage.

Because of the low Prandtl number of mercury, the thermal boundary layer on the sensor will be much larger than the hydrodynamic boundary layer, and a larger heating current will be required to achieve an adequate overheat ratio. Sensors are usually operated at a temperature about 20°C above ambient to achieve a good compromise between probe life and velocity sensitivity.

### Contamination of mercury

Probe fouling is compounded by the opacity of mercury, which prevents observation of the probe. Despite this, measurements in mercury were made as early as 1965 by Sajben, using an insulated hot wire probe. Although probe fouling was not eliminated, a calibration scheme was developed to compensate for the gradual increase of contaminants.

Probes used for mercury measurements should be kept as clean as possible to retard the onset of fouling. Hoff (1968) stored probes in closed containers of distilled water when not in use, and Gardner and Lykoudis (1971) kept probes under the surface of the mercury at all times to eliminate their passage through the contaminated free surface and to allow the fouling to reach a level beyond which no further buildup of contaminants would occur. In this way the contaminant layer on the probe remained almost constant, and because the probes were calibrated in this fouled condition, accurate velocity measurements were possible.

A major path of dirt entry into mercury is through the free surface where dust from the air settles. Yet, an enclosed test facility inhibits operational flexibility. A second entry path for contaminants is at the walls of containers and test channels where particles of dirt may dislodge.

The amount of dirt passing through the mercury-air interface can be reduced by floating a thin layer of water over the free surface of the mercury. Dust falling on the water will float to the surface and not enter the mercury. The water layer also retards oxidation. The entry of dirt from the walls of the container will be reduced if the container is made from plastic or other nonoxidizing material.

Mercury should be cleaned periodically even if every precaution is taken to keep the mercury free from contaminants. Do this by installing mercury filters in channels of the test facility for continuous filtering during testing. Otherwise, remove the mercury from the test facility and pour it by hand through filters. Hill and Sleicher (1971) cleaned mercury by bubbling oxygen continuously through it for several days, followed by repeated washing of the mercury with nitric acid. The mercury was then allowed to stand, so that the oxides could float to the surface. It was then passed through a mercury filter.

**Heat transfer in mercury**

When a heated sensor is placed in liquid metal having high thermal diffusivity, heat conduction in the fluid causes the formation of a large, round, thermal boundary layer around the sensor. This means that the heat transfer in mercury will be different from that in water, and the directional sensitivity in mercury measurements will be reduced as well.

Heat transfer in mercury is usually expressed as a relationship between the Nusselt number and the Peclet number. The Peclet number, Pe, is defined as

$$\text{Pe} = \frac{Ux}{\alpha}$$

where $x$ is a characteristic length, usually the diameter of the cylindrical sensor, and $\alpha$ is the thermal diffusivity. The Peclet number is a nondimensional fluid velocity formed as the ratio of the fluid speed and the speed at which heat diffuses through the fluid.

Several empirical relationships for the heat transfer from a cylindrical sensor in mercury have been reported by Sajben (1965). One is a heat transfer law proposed by Grosh and Cess (1958) that can be used in the Reynolds number range of $0 < \text{Re} < 60$ and is expressed as

$$\text{Nu} = 1.015\sqrt{\text{Pe}}$$

Another, proposed by Cole and Roshko (1954) for higher Reynolds numbers, is

$$\text{Nu} = \frac{2}{1.521 - \ln \text{Pe}}$$

A final relationship, by Kramers (1946), written in terms of both the Peclet number and the Prandtl number, is

$$\text{Nu} = 0.42\text{Pr}^{0.20} + 0.57\text{Pr}^{-0.33}\text{Pe}^{0.5}$$

This expression is valid over the range $10^{-2} < \text{Re} < 10^4$.

Calibration results can be graphed in terms of Nusselt number versus Reynolds number. An example of a calibration curve, by Hill and Sleicher (1971), for three hot film probes in mercury is shown in Figure 5.10, where the same data is also expressed in terms of the Sajben $X$-factor (Sajben, 1965), defined in the next section, versus the Reynolds number to obtain more consistent results.

**The calibration of probes in mercury**

Probes can be calibrated in mercury by using the techniques discussed for water calibration. Usually a calibration method requiring as little mercury as possible is chosen; for example, a common method is to use a small rotating open container of mercury over which a stationary probe is

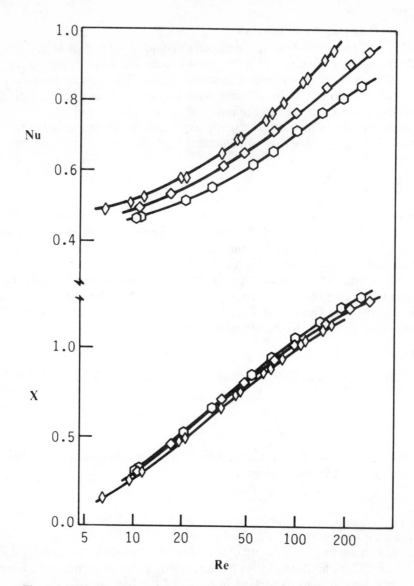

Figure 5.10. Calibration curves for three hot film probes in mercury at zero yaw angle showing the superiority of the Sajben $X$-factor over the Nusselt number if we wish to collapse the data onto a single curve. Reprinted with permission from J. C. Hill and C. A. Sleicher. Directional sensitivity of hot film sensors in liquid metals, *Rev. Sci. Instr.*, 42 (1971), 1461–1468.

mounted. The usual precautions regarding safe handling of mercury apply; for example, the open container should be covered with a thin layer of water at all times to inhibit the passage of mercury vapor into the air. Other safety measures will be discussed in greater detail later.

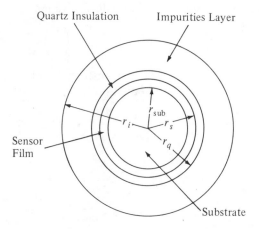

Figure 5.11. Cross section of a cylindrical film sensor surrounded by a layer of impurities, where $r_i$ is the radius of the impurities layer, $r_q$ is the radius of the quartz insulation coating, $r_s$ is the radius of the sensor film, and $r_{sub}$ is the radius of the substrate material.

To detect the slippage of mercury in a rotating tank calibration, a small permanent magnet can be floated on the surface of the mercury and its passage detected with a magnetic sensor positioned overhead. The sensor counts the number of times per second the magnet passes underneath, and this can be compared to the angular velocity of the tank (Robinson and Larsson, 1973).

No matter which calibration facility is used, the problem of inaccuracies due to probe fouling remains. Sajben (1965) was able to overcome the fouling problem with a technique, also used by Malcolm (1969) and Holroyd (1979), that compensates for both the gradual accumulation of contaminants on the sensor and the sudden increase in fouling as the probe is passed through the free surface of the mercury. The Sajben method requires the calibration of a probe in mercury while fouled with impurities, followed by a calculation to eliminate the dependence of this calibration on the thickness of the impurities layer.

If the layer of impurities is assumed to be uniform over a cylindrical sensor, its cross section will look as shown in Figure 5.11. From elementary one-dimensional conduction heat transfer through the layers on the sensor, the heat flow from the sensor is

$$q = \frac{2\pi l(T_s - T_i)}{(1/k_q)\ln(r_q/r_s) + (1/k_i)\ln(r_i/r_q)}$$

Also, convective heat transfer from the impurities layer to the fluid, based on eqn. 3.1, is

$$q = 2\pi r_i h(T_i - T_f)l$$

Combining these two equations to eliminate $T_i$, we get

$$\frac{2\pi l(T_s - T_f)}{q} = \frac{1}{r_i h} + \frac{1}{k_q} \ln\left(\frac{r_q}{r_s}\right) + \frac{1}{k_i} \ln\left(\frac{r_i}{r_q}\right)$$

The first term on the right side can be transformed into the inverse of the Nusselt number by multiplying the equation by $k_f/2$:

$$\frac{\pi l k_f(T_s - T_f)}{q} = \frac{1}{\text{Nu}} + \frac{k_f}{2}\left[\frac{1}{k_q}\ln\left(\frac{r_q}{r_s}\right) + \frac{1}{k_i}\ln\left(\frac{r_i}{r_q}\right)\right]$$

This can then be expressed as

$$D = \frac{1}{\text{Nu}} + C \tag{5.1}$$

where

$$D = \frac{\pi l k_f(T_s - T_f)}{q}$$

and

$$C = \frac{k_f}{2}\left[\frac{1}{k_q}\ln\left(\frac{r_q}{r_s}\right) + \frac{1}{k_i}\ln\left(\frac{r_i}{r_q}\right)\right] \tag{5.2}$$

The "fouling factor," $C$, depends upon the radius of the impurities layer, as does the Nusselt number. Because the cross sectional shape of the impurities layer may not be circular and, if it were, the radius may be difficult to measure, a Nusselt number based on the diameter of the quartz coating is used instead. Although this simplification alters the dimensional integrity of eqn. 5.2, it makes possible a calibration technique of great experimental usefulness.

Construct a graph of resistance vs. temperature for the sensor and use it to find the operating temperature, $T_s$, of the sensor. Since the sensor heating current is known at any flow rate, the $I_s^2 R_s$ loss for the sensor can be calculated and set equal to $q$. Measure the sensor length and calculate $D$ for any fluid speed.

Equation 5.1 for zero velocity is

$$D(0) = \frac{1}{\text{Nu}(0)} + C$$

Subtracting this from eqn. 5.1 gives

$$D(0) - D = \frac{1}{\text{Nu}(0)} - \frac{1}{\text{Nu}}$$

Note that $C$, the term depending upon the impurities layer thickness, is missing. The equation above forms the basis of the Sajben $X$-factor, defined as

$$X = D(0) - D$$

It is independent of the impurities layer thickness. Plot the mercury data as $X$ versus Pe to eliminate contamination effects.

An actual calibration is performed by testing a sensor at a variety of mercury velocities and also at zero velocity, after which $X$ is calculated for each reading. This must be done without removing the sensor from the mercury between readings in order to keep the thickness of the impurities layer constant. Then make a calibration curve of $X$ versus Pe. For velocity measurements place the probe in the test facility and measure $D(0)$. Next measure $D$ at the velocity in question and calculate $X$ for that velocity. Refer to the calibration curve to find the corresponding value of Pe, from which the mercury velocity can be found.

### Sensitivity to ambient temperature changes in mercury

A sensor used in mercury is extremely sensitive to temperature changes, and the following analysis from Malcolm (1969) illustrates the large errors possible in velocity measurements for a small change in ambient temperature.

For this analysis the $X$-factor, before the ambient temperature modifies it, is

$$X_1 = \pi l k_f \left[ \frac{\Delta T_1}{q_1(0)} - \frac{\Delta T_1}{q_1} \right]$$

where $\Delta T = T_s - T_f$, and the subscript 1 refers to conditions before the fluid temperature changes. If the fluid temperature were to change, the $X$-factor would become

$$X_2 = \pi l k_f \left[ \frac{\Delta T_2}{q_2(0)} - \frac{\Delta T_2}{q_2} \right] \tag{5.3}$$

where the subscript 2 refers to conditions after the temperature change.

Malcolm (1969) showed $D(0)$ to be a very weak function of temperature, allowing us to assume that, for small variations in fluid temperature,

$$\frac{\Delta T_1}{q_1(0)} = \frac{\Delta T_2}{q_2(0)}$$

This can be substituted into eqn. 5.3 to give

$$X_2 = \pi l k_f \left[ \frac{\Delta T_1}{q_1(0)} - \frac{\Delta T_2}{q_2} \right]$$

A person unaware of a fluid temperature change would assume the changing $X$-factor to be due to a change in velocity and would calculate an apparent $X$-factor, $X_a$:

$$X_a = \pi l k_f \left[ \frac{\Delta T_1}{q_1(0)} - \frac{\Delta T_1}{q_2} \right]$$

This can be rearranged to give

$$X_a = \pi l k_f \left[ \frac{\Delta T_1}{q_1(0)} - \left(\frac{\Delta T_2}{q_2}\right)\left(\frac{\Delta T_1}{\Delta T_2}\right) \right]$$

The relative error in temperature difference, $\epsilon_T$, is defined as

$$\epsilon_T = \frac{\Delta T_2 - \Delta T_1}{\Delta T_1}$$

and the relative error in $X$-factor, $\epsilon_X$, is defined as

$$\epsilon_X = \frac{X_a - X_2}{X_2}$$

A new variable, $Z$, is defined as

$$Z = \frac{\Delta T_1/q_1(0) - \Delta T_2/q_2}{\Delta T_1/q_1(0)}$$

The relative error in $X$-factor becomes

$$\epsilon_X = \left(\frac{1}{Z} - 1\right)\left(1 - \frac{1}{\epsilon_T + 1}\right)$$

To illustrate how a small temperature change causes a large change in $X$-factor, suppose $\epsilon_T = 0.5\%$ and a typical value for $Z$ is 0.02 (Malcolm, 1969). The relative error in $X$-factor is then about 24%. From a graph of $X$ versus Pe an error in $X$-factor of this magnitude is seen to result in a correspondingly large error in velocity.

### The directional sensitivity of probes in mercury
Because mercury is an excellent conductor of heat, the large, rounded, thermal boundary layer that forms on the sensor results in poor directional sensitivity. This was investigated by Hill and Sleicher (1971), who found that the yaw factor, $k$, defined in eqn. 2.1, was not constant with the yaw angle, $\theta$, for $\theta <45°$. Equation 2.1 can be used to represent the yaw factor at $\theta = 45°$ because the yaw factor is fairly constant at that angle. Figure 5.12 shows the variation in yaw factor with Reynolds number. This figure also shows the yaw factor to be low enough for some directional sensitivity to exist at a Reynolds number below about 200; this may be adequate for some experimental work.

The errors possible in low-velocity applications when a sensor is calibrated in parallel flow, for example, and then used in either cross or contra flow were discussed in Section 3.3. Hill and Sleicher (1971) have shown this effect to be absent for probes used in mercury.

### The effect of magnetic fields
When an electrically conducting fluid such as mercury flows through a magnetic field aligned normal to the velocity vector, electromagnetic forces

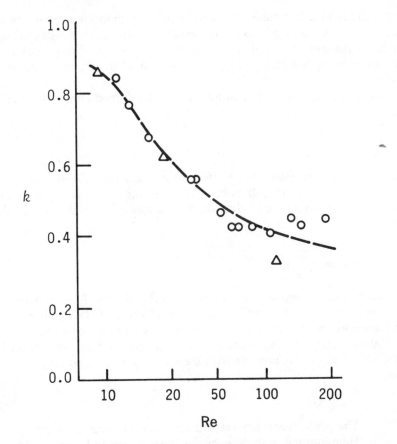

Figure 5.12. The variation in yaw factor with Reynolds number at $\theta = 45°$ for mercury. Reprinted with permission from J. C. Hill and C. A. Sleicher, Directional sensitivity of hot film sensors in liquid metals, *Rev. Sci. Instr.*, 42 (1971), 1461–1468.

alter the flow field to suppress vorticity components perpendicular to the magnetic field. If the vortices in the wake of a hot film sensor are altered, errors can result.

The vortices shed by a cylindrical sensor of infinite length have axes aligned parallel to the sensor, and a magnetic field aligned parallel to this sensor will, in theory, cause no errors. The effect of a magnetic field on vortices shed by the sensor cannot be completely eliminated in practice because actual cylindrical sensors have only moderate aspect ratios and produce vortices with orientations in many directions; wedge and conical film probes generate even more confused vortex patterns. In order to minimize the influence of magnetic fields, high-aspect-ratio cylindrical sensors aligned parallel to the magnetic field should be used, if possible.

In addition to its effect on vortices, the magnetic field also induces an electric current in the sensor when the sensor, the velocity vector, and the magnetic field are mutually perpendicular. This induced current unbalances the Wheatstone bridge, and the feedback amplifier compensates to cause an error in the output signal. Use cylindrical or wedge film probes with the sensor axis aligned parallel to the magnetic field to minimize this source of error.

The magnetic interaction parameter, $N$, for forced convection is defined as

$$N = \frac{\text{Ha}^2}{\text{Re}} \tag{5.4}$$

It is used to characterize the effect of a magnetic field on hot film measurements in mercury (Malcolm, 1970). In eqn. 5.4, Ha is the Hartmann number, which is a ratio of electromagnetic forces to viscous forces, defined as

$$\text{Ha} = \sqrt{\frac{\sigma}{\nu}} \, Bl$$

where $B$ is the magnetic flux density, $\sigma$ is the electrical conductivity of the mercury, and $l$ is the sensor length. A small value of Ha signifies minimum influence by the magnetic field on the vortex pattern behind the sensor.

Malcolm suggests that eqn. 5.4 be modified for free convection by replacing the Reynolds number with the Grashof number to give

$$N = \frac{\text{Ha}^2}{\text{Gr}}$$

### The probe frequency response in mercury measurements

Heat storage in the sensor has been shown in Section 3.8 to be the primary cause of decreased sensor frequency response in hot wire probes. Even though the hot film sensors universally used in mercury have thin films with little heat storage capacity, the mercury surrounding the sensor stores heat and significantly reduces the sensor frequency response.

Robinson and Larsson (1973) defined an approximate frequency response for low-velocity mercury as

$$\text{fr} = \frac{\pi \alpha}{4p^2}$$

where the frequency response, fr, is the frequency for which the amplitude of the output signal is decreased by 3db relative to the amplitude of the output signal under steady-state conditions, $\alpha$ is the thermal diffusivity of mercury, and $p$ is the distance from the surface of the sensor for which the mercury temperature is one-third that of the sensor surface. At higher velocities the frequency response is estimated by Robinson to be

$$\text{fr} = \frac{U^2}{4\pi\alpha}$$

This equation is invalid when the thermal wavelength becomes smaller than the dimensions of the sensor.

### Health hazards when using mercury

Those who work in the vicinity of mercury should be aware of the health dangers caused by exposure to liquid mercury or its vapor. Mercury is poisonous and can enter the body either through cuts in the skin or by inhalation of vapor when mercury is exposed to the air (Patty, 1981, pp. 1775–1789). Mercury attacks the nervous system, primarily the brain, where it is retained for a half-life on the order of years.

The U.S. Government Occupational Safety and Health Administration has determined that the maximum quantity of mercury vapor in the air that produces no toxic effect is 0.1 mg/m$^3$. Acute symptoms occur after an exposure to about 8.5 mg/m$^3$ mercury vapor in the air. When liquid mercury is injected under the skin, quantities from 40 g to 270 g can cause death.

There are two ways to reduce the possibility of contact with mercury; one is to observe safe-handling procedures and the other is to design the laboratory and the test setup for adequate containment of mercury. Of the two, safe-handling guidelines can be observed at little expense and should be practiced by all who use the laboratory.

A primary rule of safe handling is to avoid spills. Mercury can be siphoned or poured from one container to another, but a tray should be placed underneath to catch spills. A vacuum cleaner can be used to pick up droplets of spilled mercury, but a trapping container must be used between the nozzle and the blower to prevent entry of mercury into the blower housing.

Personal hygiene is important. All folds and seams in the clothing should be periodically inspected for droplets of mercury; pants cuffs and pockets should receive particular attention. Always wash your hands after each test to prevent the possibility of mercury entering the skin or being carried to your mouth; finally, do not eat in a room where experiments with mercury are being conducted. That room should also not be used as office space.

At the conclusion of a test the mercury should be returned to its containers, and the room should be cleaned. The room should be well ventilated, and the floor free from cracks that could hold small droplets of mercury. The test setup should be mounted over trays containing water or oil to stop mercury vapor from entering the air in case a spill occurs. Tables should have tops constructed with raised edges to confine spills, and, finally, a layer of water should cover the surface of all open containers of mercury.

## 5.5 Measurements in other liquids

Although hot wire anemometry is used in a variety of fluids, little may be published because researchers do not realize the pioneering nature of their work, or it may be proprietary to the company and only published internally. Therefore, measurements in only a few liquids, other than water, polymer solutions, and mercury, are described in the literature.

**Measurements in blood**

Special probes are usually required for blood flow measurements because the standard hot film probes are large enough to block the flow and are difficult to insert into a blood vessel and position accurately.

Although the primary design requirement for a blood flow probe is small size, they often have a right-angle configuration for alignment with the streamwise component of the flow. Conical and flat-surface sensors are often used. An example is the probe design of Seed and Wood (1970a), which has three film sensors; the center sensor is heated, and the two flanking sensors are used as thermometers to find the direction of the flow. This probe is mounted in the end of a hypodermic needle for easy insertion into a blood vessel. Another example is the flat-surface film probe built into the end of a hypodermic needle by Ling, Atabeck, Fry, Patel, and Janicki (1968). Bellhouse, Schultz, and Karatzas (1966) mounted their hot film probe on the end of a catheter; the entire assembly could be sterilized without damage. An example of a commercially available hot film probe mounted on a catheter is shown in Figure 5.13.

Although hot wire probes have not been used for blood flow measurements, Seed and Wood (1970) used a hot film probe having no protective coating over the sensor. Although velocity measurements were possible, fibrin deposits on the sensor were difficult to remove without damage to the thin film.

Blood can be damaged if its temperature becomes too high; therefore the temperature should not be allowed to increase more than 5°C (Seed and Thomas, 1972). At higher temperatures red-blood-cell damage can occur, and fibrin deposits will collect on the sensor. Even at low temperature differences, the blood tends to adhere to the probe, but Blick, Sabbah, and Stein (1975) reduced this by immersing a clean probe in a silicone solution, rinsing it in water, and allowing it to dry overnight before use.

Positioning of the flat-surface film probe for blood flow measurements can be difficult because the sensor must be flush with the inner surface of the opaque blood vessel. Because the exact thickness of the blood vessel is unknown, the sensor may not always be flush with the wall, and this would cause errors in shear-stress measurements. An ultrasonic positioning method was developed by Miller (1980) to overcome this problem. An ultrasonic probe was mounted on the opposite side of the blood vessel, and a sound pulse transmitted that reflected from the sensor and the inner wall of the blood vessel to measure the relative displacement of the two. The ultrasonic echoes are shown in Figure 5.14. The resolution of this technique was found to be 0.1 mm.

Probe calibration poses special problems in blood flow measurements because probes are sensitive to the hematocrit content of blood (Ling, Atabeck, Fry, Patel, and Janicki, 1968), and, consequently, blood from the same animal should be used for both the calibration and the velocity study. In addition, the amount of blood available for calibration may be limited, so the

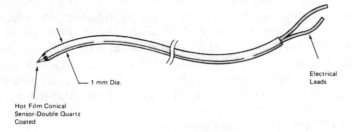

Figure 5.13. A conical hot film probe mounted on a catheter and suitable for blood flow measurements. Reprinted with permission from TSI, Inc.

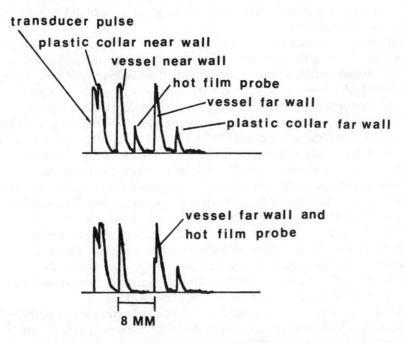

Figure 5.14. Ultrasonic echoes of aorta, collar, and hot film probe positions during in vivo experiments requiring the location of a flat surface hot film probe flush with the inner surface of the aorta of a dog. The upper trace shows the ultrasonic echo with the hot film probe in the center of the vessel lumen, and the lower trace shows the probe properly placed flush with the aortic wall. Reprinted with permission from G. E. Miller, Position sensitivity of hot-film shear probes, *J. Phys. E.: Sci. Instr.*, 13 (1980), 973–976.

calibration facility should require as little blood as possible. Furthermore, although other fluids cannot be substituted for blood during calibration, alternative fluids can be used for initial testing of the calibration facility. Ling, Atabeck, Fry, Patel, and Janicki (1968) used a glycerine-water mixture having the same kinematic viscosity as blood for this purpose.

Some interesting calibration facility designs have been developed for blood-probe calibration. A design utilizing a rotating tank into which a stationary probe is dipped was used by Seed and Wood (1969) and is similar to a design previously discussed for calibration in water. If a shaking probe calibration is desired, the probe can be vibrated in the streamwise direction while in the rotating tank (Seed and Wood, 1970). Another method is to use a closed-circuit blood tunnel. In such a design by Ling, Atabeck, Fry, Patel, and Janicki (1968), a variable-speed pump was used to move the blood through a 1.27-cm ID precision glass tube.

If a calibration of shear stress for a flat-surface probe is needed in blood, a design having two concentric cylinders, with the outer one rotating and the inner one fixed (Tillman and Haubinger, 1978), can be used. Mount the flat-surface probe flush with the inner wall and pour the blood into the space between the cylinders.

Although it is usual to allow the animal to die at the end of a blood flow experiment, every precaution is taken to preserve the life and safety of the subject in human blood flow measurements. This is done by proper electrical grounding of the equipment, sterilizing probes before use, and proper use of anticoagulants to prevent the formation of blood clots during the test.

When electronic instrumentation is connected to a human, the possibility of death by electric shock cannot be discounted. The largest electric current a human can safely experience is about 20 μA (Seed and Thomas, 1972). Above this, death could occur if ventrical fibrillation is triggered.

Instrumentation is always grounded through its wall plug, but electric shock is much less likely if the subject is also grounded. Fortunately, this is traditional practice in hospital operating rooms to prevent explosion of the anesthesia if sparks occur. Additional protection can be obtained in blood flow measurements if probes with low-resistance sensors operated at a low overheat ratio are used to insure that the voltage drop across the sensor is small. For further protection the insulation covering the sensor should have a resistance of at least 1 MΩ (Seed and Thomas, 1972).

Any foreign object introduced into the blood increases the risk of thrombus formation, which could be carried to the brain, especially if it occurs in the ascending aorta. This danger can be reduced by treating the subject with anticoagulants before testing (Seed and Thomas, 1972).

### Measurements in glycerine

Hot wire measurements in glycerine can be made with little difficulty if one is experienced in using probes in water. Although glycerine, unlike tap water, is an electrical insulator, uninsulated hot wire probes cannot be used because air dissolved in the glycerine causes corrosion of the metal sensor (Herzog and Lumley, 1978).

Dissociation of glycerine at high temperatures can be a problem, and a moderate overheat ratio must be used to prevent it. Herzog and Lumley

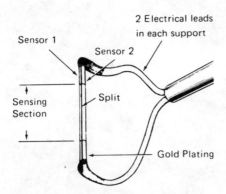

Figure 5.15. The split film cylindrical hot film probe has two metal films deposited along the length of a quartz fiber so that each film covers approximately 180° of the cylinder surface. Each support needle contains two conductors, one for each sensor. Reprinted with permission from TSI, Inc.

used a sensor temperature of about 35°C above the ambient glycerine temperature of 35°C with no dissociation occurring.

Besides problems with dissociation, glycerine has a relatively high Prandtl number compared to other liquids commonly used with hot wire anemometry, and this means that the thermal boundary layer will be much smaller than the hydrodynamic boundary layer. For this reason Herzog and Lumley found a split film sensor (Figure 5.15) to show an irregularity in the output signal when the velocity vector pointed toward the space between the two films.

### Measurements in oil

Oil is an electrical insulator, and although uninsulated hot wire sensors can be used in this liquid, there are several reasons why hot film probes should be used for all measurements. Hot wire sensors are usually made of platinum or platinum-plated tungsten wires – and platinum acts as a catalyst for the cracking of oil. When an uninsulated hot wire probe is used, dissociation of the oil at the surface of the sensor (Eckelmann, 1972) causes a viscous layer to cover the sensor. These deposits are shown on a platinum sensor in Figure 5.16. Although tungsten is not a catalyst, Figure 5.17 shows that some formations of viscous deposits are possible with this material.

Passivation of a platinum sensor also forms a hard coating, and although Eckelmann was able to remove it with a jet of methanol, the coating was sometimes swept away during a velocity measurement, causing a sudden change in sensitivity. The use of hot film probes overcomes all of these problems and gives drift-free results because the quartz insulation coating prevents contact between the sensor and the oil.

A minor consideration is that a moderate overheat ratio is needed to pre-

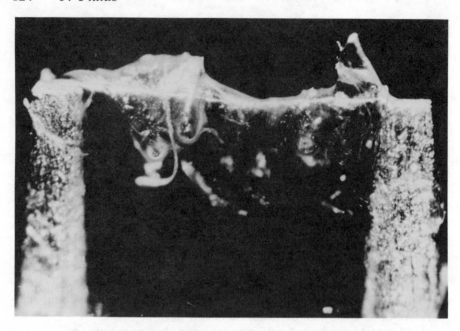

Figure 5.16. The fouling by dissociated oil of a platinum hot wire sensor used in silicone oil. Reprinted with permission from H. Eckelmann, Hot-wire and hot-film measurements in oil, *DISA Info.*, 13 (1972), 16–22.

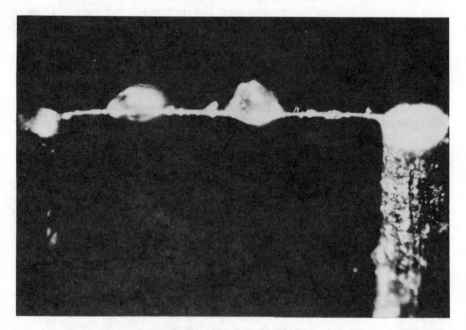

Figure 5.17. The fouling by dissociated oil of a tungsten hot wire sensor used in silicone oil. Reprinted with permission from H. Eckelmann, Hot-wire and hot-film measurements in oil, *DISA Info.*, 13 (1972), 16–22.

vent dissociation of the oil. Although the dissociation temperature for oil is about 200°C, Eckelmann chose to limit sensor temperatures to 100°C in an ambient oil temperature of 35°C.

### Measurements in luminous gas

The problems of making hot wire anemometer measurements in flames are many; the temperature is usually high enough to damage the probe, although cooled sensor probes can be used, and both the temperature and concentration of the gas can vary across the flame. In some applications small particles are carried along with the gas to quickly damage a probe. Calibration is also difficult.

One way around these difficulties is to use a gas with a low flame temperature. Wooldridge and Muzzy (1966) used 4% hydrogen and 96% nitrogen and obtained low-temperature flames that could be considered to have the properties of pure nitrogen.

# 6 TYPES OF MEASUREMENTS

For some instruments a general knowledge of their operation is enough to enable using them in any application, but this is not the case with the hot wire anemometer. For example, one experienced in using the hot wire anemometer for turbulence measurements would require a relearning period before gas mixture concentration measurements could be attempted, because specialized probes operating on different principles are required.

In this section we discuss measurements of the speed and direction of the flow as well as turbulence measurements, near-wall measurements, and shear measurements. We also discuss compressible flow measurements as well as measurements of vorticity, simultaneous velocity and temperature, gas mixture concentration, two-phase flow, and temperature fluctuations.

## 6.1 Measurements of the speed and direction of the flow

Those unfamiliar with the hot wire anemometer sometimes have the mistaken impression that one can measure the speed and direction of the flow by simply placing the probe of a hot wire anemometer in the flow and reading the speed and direction from a precalibrated front-panel meter. In reality, such a measurement requires multiple systems, calibration of each probe, a knowledge of the problems of the forward–reverse ambiguity, and the ability to accurately align the probe with the mean velocity vector.

### The forward–reverse ambiguity

With most hot wire probes one will not be able to detect a reversal of the flow direction, and, although this forward–reverse ambiguity is unimportant if a flow visualization technique has shown flow reversal not to occur, an unexpected flow reversal in the vicinity of the probe might be interpreted as a decrease in velocity followed by an increase. This is illustrated in Figure 6.1. The top graph shows a typical flow having periodic flow reversals indicated by the descending of the trace below the horizontal axis. The hot wire anemometer response shows the typical "folding" of the velocity signal that occurs with forward–reverse ambiguity.

One solution is to shield the sensor to make it sensitive to flow in only one direction. A probe of this type, patented by Neuerburg (1969), is shown in Figure 6.2. As the accompanying graph illustrates, only the forward move-

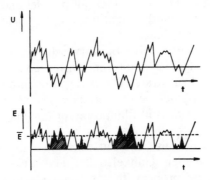

Figure 6.1. Forward–reverse ambiguity. The top graph shows free stream velocity fluctuations, with the trace below the horizontal axis indicating flow reversal. The lower graph shows the response of the typical hot wire anemometer to this flow reversal. Reprinted with permission from A. A. Gunkel, R. P. Patel, and M. E. Weber, A shielded hot-wire probe for highly turbulent flows and rapidly reversing flows, *Ind. Engr. Chem. Fund.*, 10 (1971), 627–631.

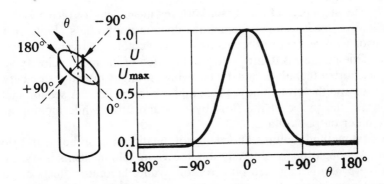

Figure 6.2. A tubular shield partially surrounding a sensor allows the fluid to pass over it in one direction only to resolve the forward–reverse ambiguity. Reprinted with permission from W. Neuerburg, Directional hot-wire probe, *DISA Info.*, 7 (1969), 30–31.

ment of the fluid is sensed when the probe is placed in a periodically reversing flow.

A second method uses two sensors: one displaced upstream from the other, and each powered separately. The heated wake from the upstream sensor is convected past the downstream sensor, causing flow disturbances and a reduction in sensitivity that can be observed by looking at the output signal from both probes; the reading of the downstream probe can then be discounted. This method was used in the hot film blood flow probe of Seed and Wood (1970a). Another approach, developed by Gupta and Srivastava (1979), and also by Mahler (1982), utilizes two separate hot wire sensors with a circular cylinder between them, all parallel and in the same plane. The downstream sensor can be identified by the wake of the cylinder superimposed on its output signal.

A split film sensor can resolve the forward–reverse ambiguity because this double-sensor design has a hot film sensor deposited on opposite sides of the same cylindrical substrate. The two sensors are powered separately. Ming Ho (1982) found a degree of thermal cross talk to exist between the sensors of a double-sensor split film probe for low-magnitude velocity fluctuation frequencies.

Forward–reverse ambiguity also causes errors in measurements when the turbulence intensity is high. Gunkel, Patel, and Weber (1971) developed a double-sensor hot wire probe with a circular shield for turbulence measurements. The circuitry used for this probe is linearized and designed to automatically discard the reading from the sensor registering the lowest velocity. This is based on the assumption that this sensor is downstream in the thermal wake of the other sensor. The shield eliminates the effect of velocity components perpendicular to the mean velocity.

### The alignment of the sensor with the mean velocity vector

Most measurement techniques depend upon one's ability to align the sensor with respect to the mean velocity vector. There are two ways to do this. One method is to measure the direction of the mean velocity vector by using another technique and, by knowing the angular relationship between the sensor and the probe body, place a probe in the flow. A second method is to angle the probe in the flow by trial and error while looking at the anemometer output voltage.

The first method requires knowledge of the direction of the mean velocity vector. Although this is generally not known in field measurements, except when the probe is mounted on a rotating vane, laboratory tests are often designed in such a way that the flow direction can be determined. The angular relationship between the sensor and the probe body must also be measured, so that the probe body can be used as a reference for alignment.

A trial-and-error method proposed by Patel (1963) can also be used. It is based on the fact that cooling is maximized with the sensor perpendicular to the mean velocity vector. For correct alignment assume the direction of the mean velocity vector and place the probe so that the sensor is perpendicular to the assumed direction. Then angle the sensor by trial and error to maximize the output voltage.

This method was automated by Bond and Porter (1967) to position a single-sensor hot wire probe normal to the mean velocity vector in a two-dimensional flow field. A servomotor rotated the probe about the probe-body axis, while a feedback control system positioned the probe at the point of maximum output voltage. It was found to be somewhat more accurate to position the sensor at the point of minimum output voltage, thus aligning the sensor parallel to the mean velocity vector; then a switch was actuated to introduce a step voltage equivalent to 90° rotation to relocate the sensor normal to the mean velocity vector.

### The measurement of the mean velocity vector

The direction and magnitude of the mean velocity vector can be found by using a triple-sensor probe or by making three consecutive measurements with a single-sensor probe. To see how this is possible, first assume a three-dimensional coordinate system with one cylindrical sensor parallel to each coordinate axis, as shown in Figure 6.3. In the figure the angles between the velocity vector and the $x$-, $y$-, and $z$-axes are given as $\alpha$, $\beta$, and $\gamma$, respectively. Use the Hinze yaw-angle relationship of eqn. 2.2 to write the effective cooling velocity for each sensor as

$$U_{\text{eff}_x}^2 = U^2(\cos^2 \alpha + k^2 \sin^2 \alpha)$$

$$U_{\text{eff}_y}^2 = U^2(\cos^2 \beta + k^2 \sin^2 \beta)$$

$$U_{\text{eff}_z}^2 = U^2(\cos^2 \gamma + k^2 \sin^2 \gamma)$$

These equations are rearranged to give the following relationship for the direction cosines of the velocity vector:

$$\cos \alpha = \frac{1}{U} \sqrt{\frac{U_{\text{eff}_x}^2 - k^2 U^2}{1 - k^2}} \tag{6.1}$$

$$\cos \beta = \frac{1}{U} \sqrt{\frac{U_{\text{eff}_y}^2 - k^2 U^2}{1 - k^2}} \tag{6.2}$$

$$\cos \gamma = \frac{1}{U} \sqrt{\frac{U_{\text{eff}_z}^2 - k^2 U^2}{1 - k^2}} \tag{6.3}$$

The equation relating the direction cosines is

$$\cos^2 \alpha + \cos^2 \beta + \cos^2 \gamma = 1 \tag{6.4}$$

Equations 6.1–6.4 are combined to give the following expression for the magnitude of the velocity vector:

$$U = \sqrt{\frac{U_{\text{eff}_x}^2 + U_{\text{eff}_y}^2 + U_{\text{eff}_z}^2}{1 + 2k^2}} \tag{6.5}$$

In practice, the magnitude of the velocity vector can be found with a triple-sensor probe powered by three separate electronics packages to measure the effective cooling velocity for each sensor simultaneously. If the yaw factor for a sensor is known and assumed to be the same for all three, the magnitude of the velocity vector can be calcuated by using eqn. 6.5. Finally, calculate the direction by using eqns. 6.1–6.3.

For an instantaneous measurement of the magnitude and direction of the

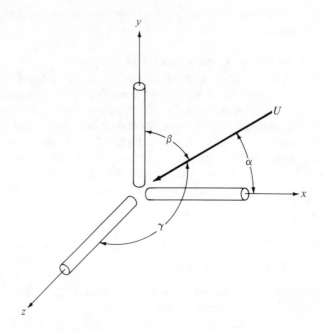

Figure 6.3. The relationship of the mean velocity vector to an idealized triple sensor hot wire probe.

velocity vector, construct a small analog computer (at this writing there are none available commercially) to solve eqns. 6.1, 6.2, 6.3, and 6.5. Or, digitize the signals from each sensor and use a digital computer to solve the equations. In either case all three channels should be "balanced" by passing the output signal from each anemometer through a separate linearizer adjusted so that all sensors have identical calibration curves.

If an instantaneous measurement is not required and a triple-sensor probe with three electronics packages is not available, an unlinearized single-sensor probe can be used to take sequential measurements with the sensor aligned with each coordinate axis. Then make a hand calculation of the direction and magnitude of the velocity vector.

For two-dimensional flow measurements a double-sensor probe can be used. Equations 6.1 and 6.2 are still valid, as is eqn. 6.4 with $\gamma = 0°$. Combine these three equations to get the following relationship for the magnitude of the velocity vector: '

$$U = \sqrt{\frac{U_{\text{eff}_x}^2 + U_{\text{eff}_y}^2}{1 + k^2}}$$

If the direction but not the magnitude of the velocity vector is needed and the quadrant in which it is located is known, a method adapted from Mojola (1974) can be used. Picture an $(x,y,z)$-coordinate system in the flow field.

## 6.2. Turbulence measurements

Place a single-sensor probe in the $(x,y)$-plane and rotate the probe axis parallel to the $z$-axis. With the sensor at one angle, record the mometer voltage and rotate the sensor to a new angle giving the same output voltage. Bisect the angle of rotation; this bisector is the projection of the velocity vector on the $(x,y)$-plane. Then place the probe in the $(y,z)$-plane, rotate the probe about an axis parallel to the $x$-axis, and use the same technique to find the projection of the velocity vector on the $(y,z)$-plane, thus completely defining the direction of the velocity vector.

## 6.2 Turbulence measurements

One of the most common uses of the hot wire anemometer is to measure turbulence. In fact, before the development of the laser Doppler velocimeter, the hot wire anemometer was the only instrument capable of making high-frequency-response measurements of velocity.

### The analysis of a sensor in a turbulent flow field

The analysis of a cylindrical sensor in a turbulent flow field is the basis for understanding the effect of small velocity fluctuations on multiple-sensor probes. In this analysis we look at the geometry of a sensor inclined in a fluctuating velocity field and then use the cosine law of cooling to derive a relationship between the anemometer output voltage and the velocity fluctuations. The following analysis is adapted from a more detailed one by Champagne and Sleicher (1967).

We begin with a cylindrical sensor of infinite length inclined in the $(s,t)$-plane of a turbulent flow field as shown in Figure 6.4. The instantaneous velocity, $U$, contains a streamwise component, $\overline{U} + u_s$, having a mean component, $\overline{U}$, as well as a component, $u_s$, that fluctuates in the streamwise direction. In addition, there are components $u_n$ and $u_t$ in the remaining orthogonal directions. The relationship between the magnitude of the instantaneous velocity vector and the magnitude of its components is

$$U^2 = (\overline{U} + u_s)^2 + u_n^2 + u_t^2$$

Angle $\theta$ lies between the normal to the sensor and the instantaneous velocity vector, $\alpha$ is the angle between the normal to the sensor and the mean velocity vector, $\beta_3$ is the angle between the sensor and the instantaneous velocity vector, and angles $\beta_1$ and $\beta_2$ are as shown.

The law of cosines for triangle $ADF$ is

$$\cos \beta_3 = \frac{(AD)^2 + (DF)^2 - (AF)^2}{2(AD)(DF)}$$

From triangle $CDF$ we have

$$\cos \beta_3 = \sin \theta$$

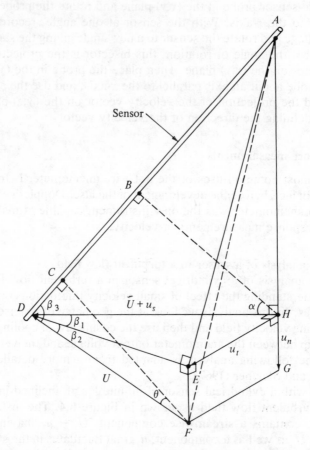

Figure 6.4. A cylindrical sensor inclined in a turbulent flow field.

Combining these two equations, we obtain

$$\sin \theta = \frac{(AD)^2 + (DF)^2 - (AF)^2}{2(AD)(DF)} \tag{6.6}$$

The sides $AD$, $DF$, and $AF$ are given, after some manipulation, by

$$AD = \frac{DH}{\sin \alpha}$$

$$DF = \frac{DH}{\cos \beta_1 \cos \beta_2}$$

$$(AF)^2 = (DH)^2 \left[ \left( \frac{1}{\tan \alpha} + \frac{\tan \beta_2}{\cos \beta_1} \right)^2 + \tan^2 \beta_1 \right]$$

Substitute these three equations into eqn. 6.6. After some manipulation, we get

$$\sin \theta = \frac{\cos \beta_1 \cos \beta_2}{2} \left( \sin \alpha - \frac{2\cos \alpha \tan \beta_2}{\cos \beta_1} - \frac{\sin \alpha \tan^2 \beta_2}{\cos^2 \beta_1} \right.$$

$$\left. - \sin \alpha \tan^2 \beta_1 + \frac{\sin \alpha}{\cos^2 \beta_1 \cos^2 \beta_2} \right) \quad (6.7)$$

For angles $\beta_1$ and $\beta_2$ we have, from Figure 6.4,

$$\sin \beta_1 = \frac{u_t}{\sqrt{(\overline{U} + u_s)^2 + u_t^2}}$$

$$\cos \beta_1 = \frac{\overline{U} + u_s}{\sqrt{(\overline{U} + u_s)^2 + u_t^2}}$$

$$\sin \beta_2 = \frac{u_n}{\sqrt{(\overline{U} + u_s)^2 + u_n^2 + u_t^2}}$$

$$\cos \beta_2 = \sqrt{\frac{(\overline{U} + u_s)^2 + u_t^2}{(\overline{U} + u_s)^2 + u_n^2 + u_t^2}}$$

Substitute these three equations into eqn. 6.7 and square to obtain

$$\sin^2 \theta = \frac{\overline{U}^2}{\overline{U}^2(1 + u_s/\overline{U})^2 + u_n^2 + u_t^2} \left\{ \left(1 + \frac{u_s}{\overline{U}}\right)^2 \sin^2 \alpha \right.$$

$$- 2 \left[ \frac{u_n}{\overline{U} + u_s} + \frac{2u_s u_n}{\overline{U}(\overline{U} + u_s)} \right.$$

$$\left. + \frac{u_s^2 u_n}{\overline{U}^2(\overline{U}^2 + 2u_s\overline{U} + u_s^2)} \right] \sin \alpha \cos \alpha \quad (6.8)$$

$$+ \left[ \frac{u_n^2}{\overline{U}^2 + 2u_s\overline{U} + u_s^2} + \frac{2u_s u_n^2}{\overline{U}(\overline{U}^2 + 2u_s\overline{U} + u_s^2)} \right.$$

$$\left. \left. + \frac{u_s^2 u_n^2}{\overline{U}^2(\overline{U}^2 + 2u_s\overline{U} + u_s^2)} \right] \cos^2 \alpha \right\}$$

The second term in the braces can be approximated by

$$-2 \left( \frac{u_n}{\overline{U}} + \frac{u_s u_n}{\overline{U}^2} \right) \sin \alpha \cos \alpha$$

the third term by

$$\left( \frac{u_n^2}{\overline{U}^2} \right) \cos^2 \alpha$$

and eqn. 6.8 becomes

$$\sin^2 \theta = \frac{(1 + u_s/\overline{U})^2 \sin^2 \alpha - 2(u_n/\overline{U} + u_s u_n/\overline{U}^2) \cdot \sin \alpha \cos \alpha + (u_n^2/\overline{U}^2) \cos^2 \alpha}{1 + 2u_s/\overline{U} + (u_s^2 + u_n^2 + u_t^2)/\overline{U}^2} \tag{6.9}$$

The denominator of eqn. 6.9 can be expressed as a binomial series of the form $(1 + x)^{-1} = 1 - x + x^2 - x^3 + \cdots$, where $x$ represents the last two terms of the denominator after expansion. The denominator then becomes

$$\frac{1}{1 + 2u_s/\overline{U} + (u_s^2 + u_n^2 + u_t^2)/\overline{U}^2}$$

$$= 1 - 2\frac{u_s}{\overline{U}} + 3\frac{u_s^2}{\overline{U}^2} - \frac{u_n^2 + u_t^2}{\overline{U}^2} - 4\frac{u_s^3}{\overline{U}^3} + 4\frac{u_s(u_n^2 + u_t^2)}{\overline{U}^3}$$

and eqn. 6.9 becomes

$$\sin^2 \theta = \left[ 1 - \frac{u_n^2 + u_t^2}{\overline{U}^2} + 2\frac{u_s(u_n^2 + u_t^2)}{\overline{U}^3} \right] \sin^2 \alpha$$

$$+ \left( \frac{u_n^2}{\overline{U}^2} - 2\frac{u_s u_n^2}{\overline{U}^3} \right) \cos^2 \alpha$$

$$- 2 \left( \frac{u_n}{\overline{U}} - \frac{u_s u_n}{\overline{U}^2} + \frac{u_s^2 u_n}{\overline{U}^3} - \frac{u_n u_t^2}{\overline{U}^3} - \frac{u_n^3}{\overline{U}^3} \right) \sin \alpha \cos \alpha$$

Because $\sin^2 \theta + \cos^2 \theta = 1$, we can write

$$\cos^2 \theta = 1 - \left[ 1 - \frac{u_n^2 + u_t^2}{\overline{U}^2} + \frac{2u_s(u_n^2 + u_t^2)}{\overline{U}^3} \right] \sin^2 \alpha$$

$$- \left( \frac{u_n^2}{\overline{U}^2} - \frac{2u_s u_n^2}{\overline{U}^3} \right) \cos^2 \alpha + 2 \left( \frac{u_n}{\overline{U}} - \frac{u_s u_n}{\overline{U}^2} + \frac{u_s^2 u_n}{\overline{U}^3} \right. \tag{6.10}$$

$$\left. - \frac{u_n u_t^2}{\overline{U}^3} - \frac{u_n^3}{\overline{U}^3} \right) \sin \alpha \cos \alpha$$

If the anemometer output voltage is processed by a linearizer, or if the variations in output voltage occur on a section of the calibration curve that can be assumed to be linear, we have that

$$E = K U_{\text{eff}}$$

where $K$ is the slope of the linearized calibration curve, and $U_{\text{eff}}$ is the effective cooling velocity, which we define by the cosine law of cooling:

$$E^2 = K^2 U^2 \cos^2 \theta$$

The anemometer output voltage, $E$, has both a mean and fluctuating part that can be expressed as $E = \overline{E} + e$; and also $U^2 = (\overline{U} + u_s)^2 + u_n^2 +$

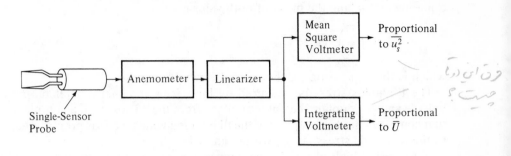

Fig. 6.5. The equipment needed to measure $\overline{u_s^2}$, the streamwise Reynolds normal stress.

$u_t^2$. Substituting eqn. 6.10 and dropping higher-order terms, we get

$$\overline{E}^2 + 2e\overline{E} = K^2(\overline{U}^2 + 2u_s\overline{U})\left[1 - \sin^2\alpha + 2\left(\frac{u_n}{\overline{U}}\right)\sin\alpha\cos\alpha\right]$$

This expression is separated into a mean part

$$\overline{\overline{E}}^2 = K^2\overline{U}^2\cos^2\alpha$$

and a fluctuating part

$$2e\overline{E} = K^2(2u_s\overline{U} - 2u_s\overline{U}\sin^2\alpha + 2u_n\overline{U}\sin\alpha\cos\alpha)$$

These last two equations are combined to eliminate $K$:

$$\frac{e}{\overline{E}} = \frac{u_s}{\overline{U}} + \frac{u_n}{\overline{U}}\tan\alpha \tag{6.11}$$

This equation will be used to analyze the effect of velocity fluctuations on a sensor.

### Using a single sensor to measure Reynolds stress

The least expensive and least complicated way to measure Reynolds stress is with a single-sensor probe; with it the Reynolds normal stresses $\overline{u_s^2}$, $\overline{u_n^2}$, and $\overline{u_t^2}$ as well as the Reynolds shear stresses $\overline{u_s u_n}$ and $\overline{u_s u_t}$ can be measured. The following analysis is based upon eqn. 6.11.

The Reynolds normal stress $\overline{u_s^2}$ can be measured by placing the sensor of a single-sensor probe perpendicular to the mean velocity vector, as shown in Figure 6.5. Equation 6.11, with $\alpha = 0°$ becomes

$$\frac{e_0}{\overline{E}} = \frac{u_s}{\overline{U}} \tag{6.12}$$

Squaring and taking the mean of both sides gives

$$\overline{u_s^2} = \left(\frac{\overline{U}}{\overline{E}}\right)^2 \overline{e_0^2} \tag{6.13}$$

which is the expression for the desired Reynolds stress.

The Reynolds stress is measured by linearizing the anemometer output voltage and ac coupling the output voltage from the linearizer to a voltmeter that computes the mean square of the fluctuating voltage; this is proportional to the Reynolds stress, as shown in eqn. 6.13.

If a measurement of turbulence intensity in a wind tunnel is needed and one can assume the turbulence to be isotropic (that is, $\overline{u_s^2} = \overline{u_n^2} = \overline{u_t^2}$), the turbulence intensity can be expressed as

$$\text{Turbulence intensity} = \frac{\sqrt{\overline{u_s^2}}}{\overline{U}} \tag{6.14}$$

This can be measured with the equipment shown in Figure 6.5, except that an rms voltmeter should be used instead of the mean square voltmeter shown. A hand calculation can be used to find the turbulence intensity.

The Reynolds shear stress, $\overline{u_s u_n}$, can be measured with a single hot wire anemometer and one inclined sensor; however, consecutive measurements must be made at two different sensor orientations. Because these measurements will not be made simultaneously, one must also assume that the magnitude of the Reynolds stress does not change with time.

In this method the sensor is mounted at $+45°$ to the mean velocity vector in the $(s,n)$-plane, and a linearizer is used to linearize the anemometer output voltage. The mean square value of the fluctuating component of voltage is found as described above, and this value is recorded along with the mean velocity and the mean output voltage. For this sensor, eqn. 6.11 applies, with $\alpha = +45°$, giving

$$e_{+45} = \frac{\overline{E}}{\overline{U}}(u_s + u_n) \tag{6.15}$$

After squaring and averaging, we get

$$\overline{e_{+45}^2} = \left(\frac{\overline{E}}{\overline{U}}\right)^2 (\overline{u_s^2} + 2\overline{u_s u_n} + \overline{u_n^2}) \tag{6.16}$$

The sensor is placed at $-45°$ in the $(s,n)$-plane, and the data taken again. For this orientation, eqn. 6.11 becomes

$$e_{-45} = \frac{\overline{E}}{\overline{U}}(u_s - u_n) \tag{6.17}$$

After squaring and averaging, we get

$$\overline{e^2_{-45}} = \left(\frac{\overline{E}}{\overline{U}}\right)^2 \overline{(u_s^2 - 2u_su_n + u_n^2)} \tag{6.18}$$

Subtract these two mean square values to obtain

$$\overline{u_su_n} = \left(\frac{\overline{U}}{2\overline{E}}\right)^2 (\overline{e^2_{+45}} - \overline{e^2_{-45}}) \tag{6.19}$$

This gives the Reynolds shear stress, $\overline{u_su_n}$, by using a single inclined sensor with measurements taken at two separate sensor orientations.

Champagne and Sleicher (1967), whose derivation used the Hinze law of cooling instead of the cosine law of cooling used here, obtained the following equation for the Reynolds shear stress:

$$\overline{u_su_n} = \left(\frac{\overline{U}}{2\overline{E}}\right)^2 \left(\frac{1 + k^2}{1 - k^2}\right) (\overline{e^2_{+45}} - \overline{e^2_{-45}})$$

where $k$ is the yaw factor. For measurement of $\overline{u_su_t}$, the above measurements can be repeated with the sensor at $+45°$ and $-45°$ in the $(s,t)$-plane.

Fugita and Kovasznay (1968) used a single sensor probe rotated slowly in the turbulent flow field by an electric motor to measure $\overline{u_s^2}$, $\overline{u_n^2}$, and $\overline{u_su_n}$. The equipment required was the same as that used in all of the methods described in this section: that is, an anemometer, linearizer, rms voltmeter, and integrating voltmeter. The output of the two voltmeters was displayed by two $(x,y)$-recorders, and the analysis based on a different law of cooling than any described in this book.

It is possible (Patel, 1963) to measure either $\overline{u_n^2}$ or $\overline{u_t^2}$ with one sensor placed in three consecutive orientations. In order to measure $\overline{u_n^2}$, first set the sensor normal to the mean velocity vector where eqn. 6.13 applies. Place the sensor first at $+45°$ and then at $-45°$ to the mean velocity vector, where eqns. 6.16 and 6.18 apply. To obtain the desired Reynolds normal stress, add and subtract the mean square values of linearized output voltage at each orientation as follows:

$$\overline{u_n^2} = \frac{\overline{U}^2}{2\overline{E}^2} (\overline{e^2_{+45}} + \overline{e^2_{-45}} - 2\overline{e_0^2})$$

The Reynolds normal stress $\overline{u_t^2}$ can be found by placing the inclined sensor in the $(s,t)$-plane.

It is possible to obtain much Reynolds stress information with a single-sensor probe, and if Reynolds stresses do not vary with time, this may be the best method to use. Otherwise, a multiple-sensor probe is needed with an increase in equipment cost by as much as a factor of five.

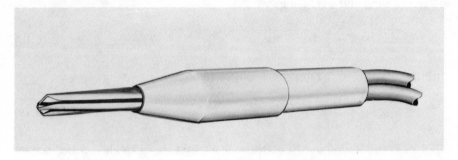

Figure 6.6. A V-array hot film probe having two hot film sensors mounted at 45° to the probe body. Reprinted with permission from Dantec Elektronik.

### ✓ Using a double-sensor probe to measure Reynolds stress

If several hot wire anemometers are available, Reynolds stresses can be measured with a double-sensor probe. An X-array or a V-array (Figure 6.6) probe is usually chosen, and a separate electronics package, linearizer, and mean square voltmeter are needed for each sensor. In addition, an integrating voltmeter is needed along with an analog computer to calculate the sum and difference of voltage signals; sum–difference computers are available commercially. In addition, the linearizers must be adjusted to give identical calibration curves for each channel.

The standard X- or V-array probe is used by first positioning it to set the sensors at ±45° to the mean velocity vector. The signals are processed from each channel with the sum–difference computer and the mean square of the voltage taken. The two linearizer output voltages are given by eqns. 6.15 and 6.17. Simultaneous adding and subtracting of these gives

$$e_{+45} + e_{-45} = 2u_s \frac{\overline{E}}{\overline{\overline{U}}}$$

$$e_{+45} - e_{-45} = 2u_n \frac{\overline{E}}{\overline{\overline{U}}}$$

$$(6.20)$$

Processing each with a mean square voltmeter gives

$$\overline{u_s^2} = \left(\frac{\overline{U}}{2\overline{\overline{E}}}\right)^2 \overline{(e_{+45} + e_{-45})^2}$$

$$\overline{u_n^2} = \left(\frac{\overline{U}}{2\overline{\overline{E}}}\right)^2 \overline{(e_{+45} - e_{-45})^2}$$

A diagram of the equipment used for this measurement is shown in Figure 6.7.

Alternatively, the mean square voltmeters can be placed ahead of the analog computer, as shown in Figure 6.8, and the two signals subtracted.

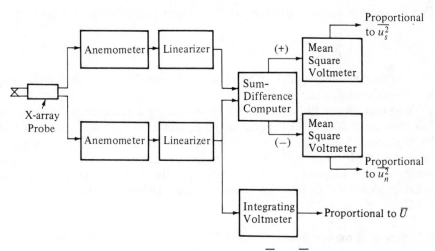

Figure 6.7. The equipment needed to measure $\overline{u_s^2}$ and $\overline{u_n^2}$, the Reynolds normal stresses, with an X-array hot wire probe.

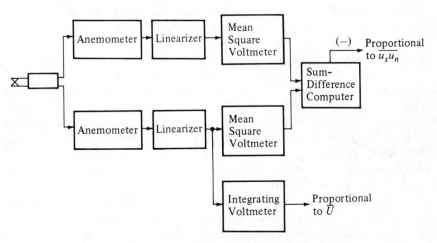

Figure 6.8. The equipment needed to measure $\overline{u_s u_n}$, the Reynolds shear stress, with an X-array hot wire probe.

Equation 6.19 applies, and the Reynolds shear stress $\overline{u_s u_n}$ can be measured. If the X-probe is placed in the $(s,t)$-plane, measurements of $\overline{u_s^2}$, $\overline{u_t^2}$, and $\overline{u_s u_t}$ can be made in a similar fashion.

A special X-array probe designed with one sensor normal to the mean velocity vector and the other at $\alpha = +45°$ can be used to measure the Reynolds normal stress $\overline{u_n^2}$ (Patel, 1963). Then eqns. 6.12 and 6.14 apply. Subtracting gives

$$e_{+45} - e_0 = u_n \frac{\overline{E}}{\overline{U}}$$

Processing this voltage difference with a mean square voltmeter gives

$$\overline{u_n^2} = \left(\frac{\overline{U}}{\overline{E}}\right)^2 \overline{(e_{+45} - e_0)^2}$$

This method should be slightly more accurate (Bradshaw, p. 169) than using a conventional X-array probe.

The preceding methods apply to flows having a turbulence intensity of less than about 20% or 30%, because the equations used have been simplified by neglecting higher-order terms that, in high-intensity turbulence, cannot be neglected. If these turbulence intensities are not exceeded, measurement errors will be no more than about 10% (Bradbury, 1976). The use of a linearizer does not usually improve accuracy (Parthasarathy and Tritton, 1963).

## 6.3 Near-wall measurements

The problem of making measurements of velocity near a wall have never been completely resolved, and the interaction between the various causes for error defy correction by analytic methods. Several empirical correction methods that have been developed are discussed below.

### The near-wall effect

Several effects cause errors when a hot wire sensor is used to make velocity measurements close to a wall. First, radiation heat loss from the sensor to a cooler wall unbalances the Wheatstone bridge, and this appears to the unsuspecting person as an increase in fluid velocity. The magnitude of this error can be estimated by moving the sensor toward the wall during zero velocity conditions while observing the change in anemometer output voltage. Dryden (1936) observed the onset of this effect in still air at a sensor-to-wall distance of about 2 mm.

If radiation were the only effect present, errors caused by the nearness of a wall could be corrected with the above method. This is not possible, because the hydrodynamic interaction between the wall, sensor, and support needles causes additional errors that are a function of the sensor-to-wall distance, support needle orientation, and fluid velocity.

The near-wall effect is illustrated graphically in Figure 6.9. Notice that the uncorrected velocity readings reach a minimum and then increase as the sensor approaches the wall. The experimental data are illustrated along with a theoretical curve based on the assumption that the universal velocity profile in the form

$$\frac{U}{v_*} = 5.6 \ln\left(\frac{yv_*}{v}\right) + 4.5$$

holds. In this equation, $v_*$ is the friction velocity. This figure shows the necessity for corrections to the data in air if the sensor is placed somewhat less than 1 mm from the wall.

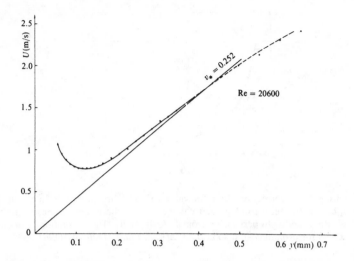

Figure 6.9. The effect of a nearby wall on velocity measurements. The solid line is the actual velocity, and the data points indicate the measured velocity. In this graph, $y$ is the distance between the wall and the sensor. Reprinted with permission from S. Oka and Z. Kostic, Influence of wall proximity on hot-wire velocity measurements, *DISA Info.*, 13 (1972), 29–33.

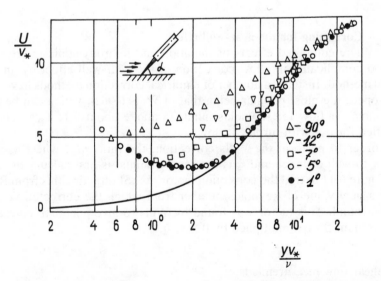

Figure 6.10. The influence of support needle inclination on velocity measurements near a wall. Reprinted with permission from A. F. Polykov and S. A. Shindin, Peculiarities of hot-wire measurements of mean velocity and temperature in the wall vicinity, *Lett. Heat Mass Trans.*, 5 (1978), 53–58.

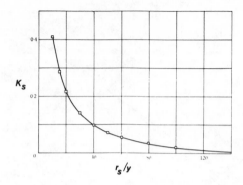

Figure 6.11. A correction curve for values of $Re^{0.45}$ for proximity to a wall. In this graph, $r_s$ is the sensor radius, and $y$ is the distance between the sensor and the wall. Reprinted with permission from J. A. B. Wills, The correction of hot-wire readings for proximity to a solid boundary, *J. Fluid Mech.*, 12 (1962), 388–396.

The effect of support needles and the probe body in near-wall measurements was investigated by Polyakov and Shindin (1978) by placing a standard hot wire probe close to a wall with the support needles and probe body at angles of from 1° to 90° to the wall. The results are shown in Figure 6.10 and indicate that even though the near-wall effect is significant at any probe body orientation, minimizing the angle between the probe and the wall tends to reduce errors, at least at some sensor-to-wall distances.

### Correcting for the near-wall effect

The combined effects of radiation and hydrodynamic interference are too complicated to allow correction of the near-wall effect by an analytical method. Instead, a number of empirical correction methods have been developed, the most popular being that of Wills (1962), which can be used in laminar air flow for Reynolds numbers above about 0.1; it can also be modified for use with turbulent flows.

In order to perform the Wills correction, the ratio $r_s/y$, where $y$ is the sensor-to-wall distance and $r_s$ is the sensor radius, is first calculated. Then use Figure 6.11 to find the correction factor, $K_s$. Subtracting this from $Re^{0.45}$ gives a new value of Reynolds number from which the corrected velocity can be found. This factor was found by Wills to give adequate correction for turbulent flow if multiplied by $0.5 \pm 0.1$.

### 6.4 Shear flow measurements

Shear flows are important not only because they can be measured by using hot wire anemometry but also because an undetected shear flow can give errors in velocity measurements.

**Velocity measurements in shear flows**

If a cylindrical sensor is calibrated in a uniform flow field and placed in a shear flow so that one end is cooled more than the other, the velocity recorded by the anemometer will be less than the average velocity experienced by the sensor. Gessner and Moller (1971) investigated this phenomenon and, in doing so, also found that turbulence measurements in shear flow give unrealistically low results. The problem was resolved for turbulence measurements either by calibrating the sensor in a shear flow or by applying a correction formula to the data. In addition, errors in shear flow measurements were found to be slightly larger for larger-aspect-ratio sensors and for those having higher overheat ratios, at least in the overheat ratio range from 0.4 to 0.8. In addition, a tungsten-wire sensor was found to give slightly less error than a platinum-wire sensor.

An infrared microscope was used by Gessner and Moller to measure the sensor temperature profile in a shear flow. Skewing of the sensor temperature profile in shear flow is shown in Figure 3.5 and can be compared to the temperature profile for uniform flow that accompanies it.

Steady-state velocity measurements can be corrected for the presence of a shear flow that causes a mean velocity gradient along the sensor by using the correction curves for platinum- and tungsten-wire sensors shown in Figures 6.12 and 6.13, respectively. The Nusselt number in these graphs is based on the assumption of negligible heat loss from the sensor to the support needles and is defined by eqn. 3.11. Nu is the Nusselt number calculated with the shear flow present, and $Nu_o$ is the Nusselt number with the probe in uniform flow. For constant temperature operation at a fixed overheat ratio, eqn. 3.11 shows that

$$\frac{Nu_o}{Nu} = \frac{I_{s_o}^2}{I_s^2}$$

The quantity $S$ is a shear factor defined as

$$S = \frac{\Delta U/U_{cl}}{l/d}$$

where $\Delta U$ is the difference in velocity seen by the two ends of the sensor, $U_{cl}$ is the velocity at the centerline of the sensor, and $l/d$ is the sensor aspect ratio. Gessner and Moller (1971) have developed similar correction curves for turbulent flow.

If velocity profile measurements in a shear flow are needed, a single-sensor probe can be placed at various locations and aligned to eliminate any velocity gradient on the sensor. For measurements close to a wall, near-wall effects may cause difficulties. If several hot wire anemometers are available, a rake of probes can be built for simultaneous measurements at many points on the velocity profile. It is good practice to compare the results from the rake of probes with the results taken with a single-sensor probe placed at various

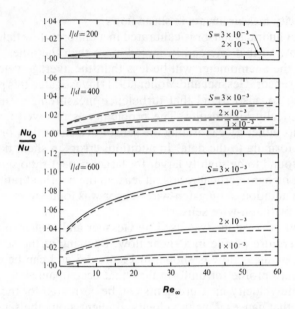

Figure 6.12. The steady-state response correction curves for platinum wires having a mean velocity gradient along their length. Reprinted with permission from F. B. Gessner and G. L. Moller, Response behavior of hot wires in shear flows, *J. Fluid Mech.*, 47 (1971), 449–468.

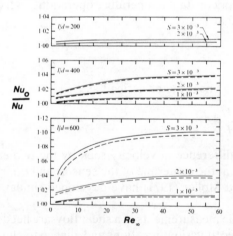

Figure 6.13. The steady-state response correction curves for tungsten wires having a mean velocity gradient along their length. Reprinted with permission from F. B. Gessner and G. L. Moller, Response behavior of hot wires in shear flow, *J. Fluid Mech.*, 47 (1971), 449–468.

locations to assure that interference between individual sensors on the rake of probes is absent.

A probe placed in a free shear flow has been found by Hussain and Zaman (1978) to cause single-frequency instabilities in the flow that are similar to edge tones. These may be sensed by the probe and give erroneous readings. Also, Hartmann (1982) examined hot wire sensors under a microscope and found them to "whirl" when placed in a shear flow; this gave a spurious sine wave superimposed on the output signal.

### Wall shear-stress measurements

A mechanical way of measuring the wall shear stress is to measure the drag force on a floating surface mounted flush with a wall, but this method is not accurate when a moderate pressure gradient is present (Liepmann and Skinner, 1954); hot wire anemometry methods do not suffer from this problem. Although early hot wire probes for wall shear-stress measurements consisted of a heated wire partially imbedded in a wall, the flat-surface hot film probe is now used instead.

Although flush-mounted hot wire probes are not manufactured commercially, one can be made by using a technique developed by Liepmann and Skinner, who glued a 0.013-mm-diameter platinum wire in a groove on an ebonite plug. The plug was mounted flush in the wall of the test section of a wind tunnel. Poor results were obtained if the groove was omitted and the wire sensor glued directly to the surface of the ebonite.

The Nusselt number for heat transfer between a surface-mounted sensor and a fluid was found by Ludwieg (1950) to be proportional to the one-third power of the shear stress. This heat transfer law was expressed by Geremia (1972) as

$$E_b^2 = A + B\tau_w^{1/3}$$

where $\tau_w$ is the shear stress at the wall, and $A$ and $B$ are constants.

A flat-surface hot film probe can be calibrated for shear-stress measurements in a variety of ways. For shear-stress measurements in air, Liepmann and Skinner (1954) used a mechanical floating-element shear-stress meter to calibrate their flat-surface film probes. For water measurements Ling, Atabeck, Fry, Patel, and Janicki (1968) used a piston to oscillate the liquid in a vertical tube; the shear stress at the maximum velocity point of each cycle was used for calibration, and the anemometer output voltage was compared to theoretical values of wall shear stress in an oscillating flow. Tillmann and Haubinger (1978) used two concentric cylinders – the inner one stationary and containing a flush-mounted flat-surface hot film probe, and the outer cylinder rotating with constant angular velocity. Because the wall shear stress for laminar Couette flow of the liquid between the cylinders is well known, the probe is easily calibrated.

## 6.5 Vorticity measurements

There are several types of vorticity measurements possible with the hot wire anemometer. One is the instantaneous measurement of the streamwise and transverse components of vorticity at a point, and another is the measurement of spatial vorticity averaged over an area in the flow field.

### The definition of vorticity

There are two definitions of vorticity that apply to measurements using hot wire and hot film techniques: the vorticity at a point, and the spatial vorticity.

Vorticity, $\boldsymbol{\omega}$, at a point in the flow is defined by

$$\boldsymbol{\omega} = \nabla \times \mathbf{U}$$

If the velocity vector is assumed to have a large mean component $\overline{U}$ in the $x$-direction and fluctuating components $u_x$, $u_y$, and $u_z$ in the $x$-, $y$-, and $z$-directions, respectively, the streamwise component of vorticity is

$$\omega_x = \frac{\partial u_z}{\partial y} - \frac{\partial u_y}{\partial z}$$

and the two transverse components of vorticity are

$$\omega_y = \frac{\partial u_x}{\partial z} - \frac{\partial u_z}{\partial x}$$

and

$$\omega_z = \frac{\partial u_y}{\partial x} - \frac{\partial u_x}{\partial y} \tag{6.21}$$

The spatial vorticity, $\overline{\omega}$, can be found by applying Stokes's theorem to the definition of circulation. Circulation, $\Gamma$, is defined as the integral of the inner product of the velocity vector and the path vector about a closed curve, $C$, or

$$\Gamma = \oint_C \mathbf{U} \cdot d\mathbf{l}$$

Stokes's theorem states that

$$\oint_C \mathbf{U} \cdot d\mathbf{l} = \iint_A \boldsymbol{\omega} \cdot d\mathbf{A}$$

where $\mathbf{A}$ is the area enclosed by curve $C$. Rearranging gives

$$\overline{\omega} = \frac{1}{A} \oint_C \mathbf{U} \cdot d\mathbf{l} \tag{6.22}$$

Both of these vorticity concepts will be used in the experimental methods to be described.

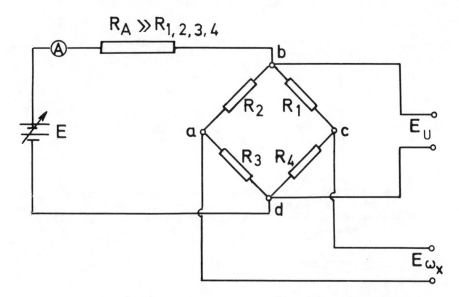

Figure 6.14. A schematic diagram of a constant current anemometer for streamwise vorticity measurements. Each resistor represents a hot wire sensor. Reprinted with permission from E. G. Kastrinakis, H. Eckelmann, and W. W. Willmarth, Influence of the flow velocity on a Kovasznay type vorticity probe, *Rev. Sci. Instr.*, 50 (1979), 759–767.

**The measurement of the streamwise component of vorticity**

The streamwise component of vorticity can be measured with the Kovasznay vorticity probe illustrated in Figure 2.27. It has four sensors of equal length set at 45° to the support needles to form a tetrahedron and is aligned in the flow with the support needles parallel to the mean velocity vector. The sensors, all of equal electrical resistance, make up the four resistors of a Wheatstone bridge circuit operated in the constant current mode. The schematic diagram of the circuit is shown in Figure 6.14. Notice that a signal proportional to $\overline{U} + u_x$ is read out at points *ab*, and a signal proportional to $\omega_x$ is read out at points *cd*, allowing simultaneous measurements of these two parameters.

With the velocity vector parallel to the support needles, all sensors are cooled equally, and the velocity is recorded at points *ab* in the schematic diagram. Because the bridge is in balance, there is no potential difference at points *cd*. To understand the sensitivity of the probe to vorticity, look end-on at the probe so that the tips of the support needles are pointing toward you. The probe is seen as shown in Figure 6.15. The letter *S* in this figure represents the tips of shorter support needles, and the letter *L* represents the tips of longer support needles. The sensors are numbered to correspond to the resistance nomenclature on the schematic diagram.

Assume the fluid to be moving toward the tips of the support needles with

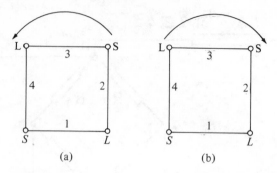

Figure 6.15. End-on view of a Kovasznay vorticity probe illustrating the effect of (a) counterclockwise vorticity and (b) clockwise vorticity. Each sensor is mounted at 45° to the support needles, and shorter and longer support needles are labeled S and L, respectively.

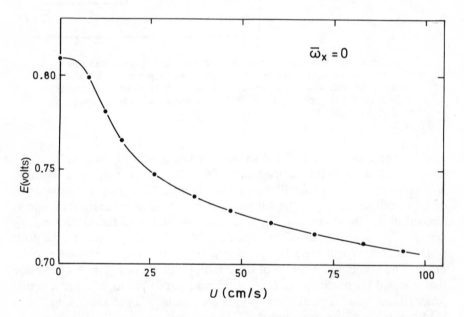

Figure 6.16. The calibration curve for a vorticity probe experiencing a streamwise mean velocity. Reprinted with permission from E. G. Kastrinakis, H. Eckelmann and W. W. Willmarth, Influence of the flow velocity on a Kovasznay type vorticity probe, *Rev. Sci. Instr.*, 50 (1979), 759–767.

a streamwise component of vorticity superimposed in the counterclockwise direction, as shown in Figure 6.15a. For sensors 1 and 3 the instantaneous velocity vector is tilted slightly, decreasing the yaw angle and cooling these sensors more. In contrast, the yaw angle of sensors 2 and 4 is increased to give less cooling.

The schematic diagram (Figure 6.14) shows the voltage drop to be less

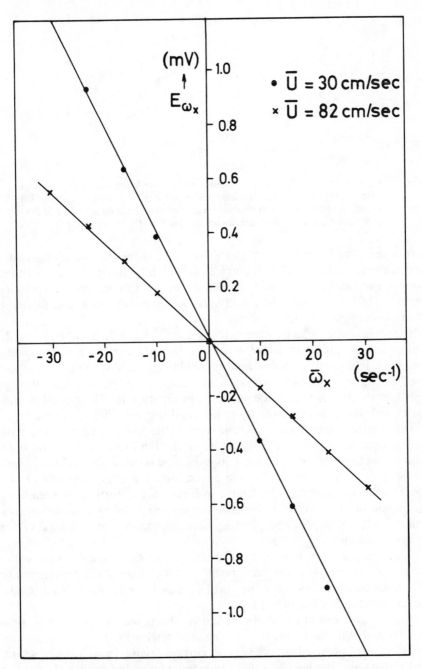

Figure 6.17. The calibration curve of a vorticity probe for the streamwise component of vorticity for two values of mean velocity. Reprinted with permission from E. G. Kastrinakis, H. Eckelmann, and W. W. Willmarth, Influence of the flow velocity on a Kovasznay type vorticity probe, *Rev. Sci. Instr.*, 50 (1979), 759–767.

Less error

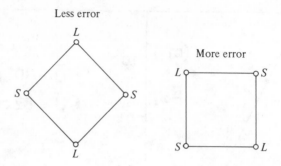

Figure 6.18. As with all multiple-sensor probes, the Kovasznay vorticity probe is susceptible to thermal interference between sensors for some orientations. The preferred orientation for minimum thermal interference is on the left, and the shorter and longer support needles are labeled S and L, respectively.

across resistors $R_1$ and $R_3$ and greater across resistors $R_2$ and $R_4$ for counterclockwise vorticity. Thus, point $c$ is at a higher electrical potential than point $d$. For clockwise vorticity, shown in Figure 6.15b, the opposite polarity occurs. Thus, the system is sensitive to both the magnitude and direction of the streamwise vorticity vector.

The streamwise velocity calibration is done separately from the vorticity calibration. A calibration for $\overline{U} + u_x$ is made in standard fashion by placing the probe in a low-turbulence wind tunnel in laminar flow with the support needles aligned with the mean velocity vector. A graph of output voltage versus resistance is then made as shown in Figure 6.16. The typical nonlinearity and decrease in sensitivity at high and low velocities can be seen.

The vorticity calibration is made by rotating the probe about its streamwise axis at constant angular velocity. Because the fluid rotates as a solid body with respect to the probe, the vorticity measured is twice the angular velocity of the probe. A typical calibration curve for streamwise vorticity at two values of mean velocity is illustrated in Figure 6.17. Vorticity is seen to be a function of mean velocity, because the sensitivity of each sensor is nonlinear, so this probe will be more sensitive to vorticity at midrange values of mean velocity.

The streamwise vorticity data is corrected by using the calibration curve for mean velocity. This can be done at the conclusion of the experiment or with a digital computer while the vorticity data is being taken (Kastrinakis, Eckelmann, and Willmarth, 1979).

Some drawbacks of this probe design are discussed by Kastrinakis, Eckelmann and Willmarth (1979). For some roll angles the probe is more sensitive to errors caused by convection currents rising from a lower sensor and influencing the heat transfer characteristics of the sensor above. For example, for the probe orientation shown in Figure 6.18, there is more error in the orientation shown at $b$. A second difficulty occurs in turbulent flow where random fluctuations in $u_y$ and $u_z$ cause substantial errors in the vorticity signal. Finally, although the four sensors are ideally of the same length and resistance, this is never the case for an actual probe and the resulting bridge unbalance will cause a voltage difference at points $cd$ with no vorticity present.

An attempt to overcome some of the above problems was made by Vukoslavcevic and Wallace (1981) with a vorticity probe of the Kovasznay type having eight support needles – two for each sensor – with each sensor powered individually by an independent constant temperature electronics package.

### The measurement of the transverse component of vorticity

The measurement of the transverse component of vorticity is based on its definition (Foss, 1978, 1981). For example, one transverse vorticity component is given by eqn. 6.21. In order to measure this component, we measure each term separately and subtract. To measure the second term, $\partial u_x/\partial y$, two sensors are placed side by side in the $(y,z)$-plane and perpendicular to the mean velocity vector. Each sensor measures $\overline{U} + u_x$ at its location. Subtracting instantaneous values of velocity measured by each sensor and dividing by the distance between the sensors gives

$$\frac{\partial u_x}{\partial y} \approx \frac{(\overline{U} + u_x)_2 - (\overline{U} + u_x)_1}{\Delta y} \tag{6.23}$$

The first term of eqn. 6.21, $\partial u_y/\partial x$, can be measured by placing an X-array probe in the flow with the sensors at 45° to the mean velocity vector and in the $(x,z)$-plane. The sensors together measure the instantaneous value of $u_y$. This is done by subtracting the signals from the two sensors of the X-array probe, as can be seen by rearranging eqn. 6.20 to give

$$u_y = \frac{\overline{U}}{2\overline{E}} (e_{+45} - e_{-45})$$

Subtracting the value of $u_y$ at two different times allows the calculation of

$$\frac{\partial u_y}{\partial t} \approx \frac{u_y(t_2) - u_y(t_1)}{t_2 - t_1} \tag{6.24}$$

and the substantial derivative of velocity is

$$\frac{\partial u_y}{\partial t} = u_x \frac{\partial u_y}{\partial x} + u_y \frac{\partial u_y}{\partial y} + u_z \frac{\partial u_y}{\partial z} + \frac{\partial u_y}{\partial t}$$

Then if $\partial u_y/\partial t \approx 0$ and $u_x \gg u_y, u_z$, we get

$$\frac{\partial u_y}{\partial t} = -u_x \frac{\partial u_y}{\partial x} \tag{6.25}$$

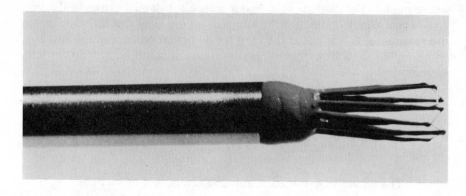

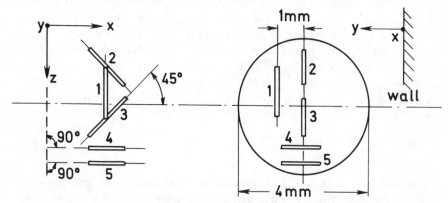

Figure 6.19. A five-sensor cylindrical hot film probe designed to measure both transverse vorticity components. The drawing shows the sensor arrangement. Reprinted with permission from H. Eckelmann, S. G. Nychas, R. S. Brodkey, and J. M. Wallace, Vorticity and turbulence production in pattern recognized turbulent flow structures, *Phys. Fluids,* 20 (1977), S225–S230.

and eqn. 6.24 becomes

$$\frac{\partial u_y}{\partial x} = -\frac{1}{u_x}\left[\frac{u_y(t_2) - u_y(t_1)}{t_2 - t_1}\right]$$

Subtracting eqn. 6.25 from eqn. 6.23 gives $\omega_z$.

A scheme allowing measurement of both transverse vorticity components (Eckelmann, Nychas, Brodkey, and Wallace, 1977) utilizes the five-sensor hot film probe pictured in Figure 6.19. As can be seen, sensor 1 lies in the $(x,z)$-plane, is perpendicular to the mean velocity vector, and measures $\overline{U} + u_x$. Beside it are sensors 2 and 3, which form a V-array and measure $U$, $u_x$, and $u_z$. These three sensors make up a parallel array allowing the measurement of $\partial u_x/\partial y$ as described above. The V-array alone measures $\partial u_y/\partial x$, and thus $\omega_z$ can be calculated. Below these three sensors are sensors 4 and 5, which form an X-array measuring $U$, $u_x$, and $u_y$. The V-array can be used

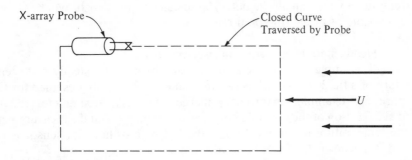

Figure 6.20. Traversing a closed curve with an X-array probe to measure spatial vorticity.

with Taylor's hypothesis to obtain $\partial u_z/\partial x$, and $\partial u_x/\partial z$ is obtained by comparing $u_z$ from both the V-array and X-array and using an equation analogous to eqn. 6.23. A subtraction gives $\omega_y$.

In contrast to these methods, the measurement of spatial vorticity, to be discussed in the next section, does not require special probes or computerized data-analysis equipment.

### The measurement of spatial vorticity

In this technique, used by Francis, Kennedy, and Butler (1978), an X-array probe is traversed around a closed curve in a region, within which the spatial vorticity measurement is required, as shown in Figure 6.20. An $(x,y)$-recorder can be used to graph the component of velocity normal to the curve versus the curve length. Integrate this graph with a planimeter or by "counting squares" and use eqn. 6.22 to calculate the average spatial vorticity. Because the closed curve can be drawn in any plane, average values of either streamwise or transverse vorticity can be measured. There is, of course, no restriction on the shape of the closed curve; use of an $(x,y)$-traverse mechanism might dictate a rectangular curve, but, despite curve shape, the two sensors of the X-array probe must always have the same angular relationship to the mean velocity vector.

If a measurement of the variation in average spatial vorticity within the closed curve is needed, more traverses can be made to subdivide the enclosed area into smaller areas. Of course, each smaller area must be large compared to the size of the sensor array. All data analysis, including integration, can be performed with a digital computer (Keesee, Francis, and Lang, 1979).

### 6.6 Temperature measurements

In Chapter 3 an expression was derived for the sensitivity of a heated sensor, which showed that the sensitivity responds primarily to fluid temperature at low overheat ratio. In this section a hot wire anemometer in the constant current mode with a small-diameter, large-aspect-ratio sensor will be shown to be capable of measuring fluid temperature.

### Steady-state temperature measurements

If a simple and inexpensive measurement of the steady-state temperature of a fluid is needed, use an instrument specifically designed for the purpose; for example, an ordinary mercury-in-glass thermometer can be used at a fraction of the cost of a hot wire anemometer. But if a measurement of the temperature at the location of the hot wire or hot film sensor is required, the sensor itself can be used.

Steady-state temperature measurements are made by first constructing a graph of sensor resistance versus fluid temperature; when a temperature measurement is needed, measure the sensor resistance and use the graph to find the temperature.

For continuous measurement of steady-state fluid temperature, one may be tempted to use the anemometer in either the constant current or constant temperature mode at a low overheat ratio. Although either mode allows continuous temperature measurements, drift can be a problem, because at low overheat ratio neither the amplifier in the constant temperature anemometer nor the constant current source in the constant current anemometer have good stability. In addition, thermoelectric forces in the probe contribute to drift.

### The measurement of temperature fluctuations

A reduction in sensor temperature was shown in Section 3.6 to cause a decrease in velocity sensitivity for either constant temperature or constant current operation. For this reason fluctuating temperature measurements are made by using a low value of overheat ratio. Yet a compromise is necessary; the overheat ratio must be as high as possible for good system stability without introducing appreciable velocity sensitivity. Either constant current or constant temperature operation can be used for fluctuating temperature measurements, but constant current operation is often chosen for improved stability. Some commercially available constant temperature anemometers can be operated in the constant current mode to allow measurement of temperature fluctuations.

A novel constant current anemometer, developed by Townsend (1951) for temperature measurements, used two sensors, one hot and the other cooler, each in an arm of a Wheatstone bridge circuit. The cooler sensor, still somewhat velocity sensitive because of the overheat ratio used, was velocity compensated by the hotter sensor to achieve a circuit that was sensitive only to temperature. This system is unusual because it is the reverse of the present practice of temperature compensating a velocity sensor with a second sensor operated at a low overheat ratio in the Wheatstone bridge of a constant temperature anemometer.

**The frequency response of temperature sensors**

The frequency response of a temperature sensor depends upon both the ability of the sensor to store heat within itself and the heat storage capability of the support needles. In the following analysis the sensor is assumed to be cylindrical in shape and of infinite length; then support-needle effects need not be considered. In addition, the sensor can be assumed to be located in one arm of the Wheatstone bridge of a constant current anemometer where the voltage drop across the Wheatstone bridge is proportional to the sensor resistance.

The heat balance equation for a cylindrical sensor of infinite length is given by eqn. 3.3. If we assume negligible radiation heat transfer and a constant temperature profile along the sensor, this equation takes the form of eqn. 3.75. For constant current operation this becomes

$$\frac{\partial R_s}{\partial t} - \left( \frac{I^2 \alpha R_o - \pi dhl}{\rho c Al} \right) R_s - \frac{\pi dh R_o}{\rho c A} (1 - \alpha T_o) = \frac{\pi dh \alpha R_o}{\rho c A} T_f$$

$$(6.26)$$

Next, both the sensor resistance and the fluid temperature are assumed to be composed of the sum of a steady-state and fluctuating part, or

$$R_s = \overline{R}_s + r_s \tag{6.27}$$

$$T_f = \overline{T}_f + \theta_f \tag{6.28}$$

where $\overline{R}_s$ and $\overline{T}_f$ are the mean sensor resistance and fluid temperature, respectively, and $r_s$ and $\theta_f$ are the fluctuating sensor resistance and fluid temperature, respectively. Two differential equations are formed – one by substituting steady-state values into eqn. 6.26, and the other by substituting eqns. 6.27 and 6.28 into eqn. 6.26. We subtract to find the differential equation for the fluctuating part to be

$$\frac{dr_s}{dt} - \left( \frac{I^2 \alpha R_o - \pi dhl}{\rho c Al} \right) r_s = \frac{\pi dh \alpha R_o}{\rho c A} \theta_f$$

The time constant for this equation is

$$\tau = \frac{\rho c Al}{\pi dhl - I^2 \alpha R_o} \tag{6.29}$$

Notice that the time constant can be reduced by decreasing the diameter of the sensor.

For the preceding analysis to be valid, the sensor aspect ratio should be large enough to make heat loss to the support needles insignificant. Bremhorst, Krebs, and Gilmore (1977) suggest that convective heat loss to support needles should be no more than about 5% of the total sensor heat loss, and that temperature sensors from 0.5-$\mu$m- to 2.0-$\mu$m-diameter platinum–iridium

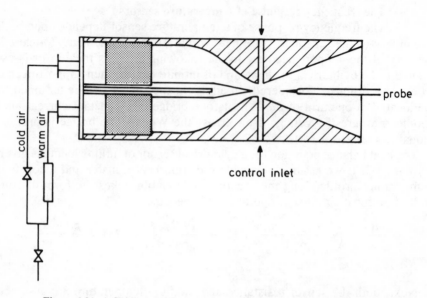

Figure 6.21. A fluidic amplifier used to oscillate a bitemperature jet across a hot wire sensor to measure its frequency response to temperature changes. Reprinted with permission from H. Fiedler, On data acquisition in heated turbulent flows, in *Proc. Dyn. Flow Conf.*, pp. 81–100, Skovlunde, Denmark, 1978.

or platinum–rhodium wire should have aspect ratios of no less than about 600.

There are several other techniques intended to subject a sensor to a fluctuating temperature field. One technique, developed by Smits, Perry, and Hoffman (1978), uses an open jet wind tunnel having a splitter plate parallel to the axis of the tunnel nozzle. This splitter plate separates the air into two streams of equal velocity, one of which is heated, and the probe is traversed between the two streams.

In another technique, developed by Fiedler (1978) and illustrated in Figure 6.21, hot and cold air separated by a splitter plate are used as the power jet of a fluidic amplifier. The power jet is made to oscillate across a sensor by means of the amplifier control jets.

A method developed by Smits, Perry, and Hoffman (1978) used a spotlight focused on the temperature sensor by an external lens. Because an incandescent spotlight cannot be turned on and off quickly, a motor-driven slotted disc was placed between the sensor and the spotlight to interrupt the light beam. Even so, at high frequencies, lack of adequate collimation can cause the light intensity to vary sinusoidally. Because there is windage associated with the slotted disc, a clear sheet of plastic was placed between the disc and the sensor to shield it from air currents. An advantage of this system is the ease with which parts of the probe can be illuminated by masking the

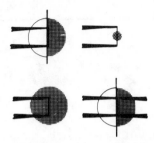

Figure 6.22. The illumination of various combinations of the sensor and support needles with a light beam to measure the frequency response of a hot wire probe to temperature changes. The dark areas in the drawing represent the light beam. Reprinted with permission from H. Fiedler, On data acquisition in heated turbulent flows, in *Proc. Dyn. Flow Conf.*, pp. 81–100, Skovlunde, Denmark, 1978.

light beam. A similar technique using a laser to heat the sensor was developed by Fiedler (1978). Here, the light beam was masked to illuminate various parts of the sensor and support needles, as shown in Figure 6.22.

Thermal white noise was used by Schacher and Fairall (1979) to measure the frequency response of a temperature sensor. Turbulent temperature fluctuations were transported downstream and past the temperature sensor, and a spectrum analyzer was used to determine the frequency response of the temperature sensor.

Another technique, developed by Hojstrup, Rasmussen, and Larsen (1976, 1977), used a cavity that resonated when air passed through it to cause pressure waves that generate temperature fluctuations. The fluctuating temperature field formed in this way was adjustable in frequency from 2 Hz to 10 kHz.

The relative frequency response between a clean hot wire probe and a salt-encrusted one was measured by Fairall and Schacher (1977) by mounting them side by side and close together inside a styrofoam-lined box containing a loudspeaker.

In a final technique, by Antonia, Browne, and Chambers (1981), a sensor upstream from the temperature sensor was heated for 1-$\mu$s duration, and the temperature spot was convected downstream to pass over the temperature sensor. The two sensors were normal to each other and the mean velocity vector and were placed 0.5 mm apart.

### Probes for temperature measurements

A standard hot wire probe can be used to measure temperature, and it is probably ideal for steady-state temperature measurements because it is rugged and available. But for fluctuating temperature measurements, a probe having a small sensor time constant is needed.

There is little disagreement between manufacturers about proper probe

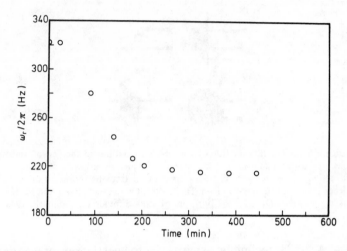

Figure 6.23. The variation of sensor rolloff frequency, ω , for a contaminated temperature sensor. Reprinted with permission from A. J. Smits, A. E. Perry and P. H. Hoffmann, The response to temperature fluctuations of a constant-current hot-wire anemometer, *J. Phys. E.: Sci. Instr.*, 11 (1978), 909–914.

design for fluctuating temperature measurements. Specifications from catalogs show the materials and dimensions to be virtually identical.

Because fluctuating temperature can only be measured accurately at frequencies below that dictated by the sensor time constant, sensor dimensions that influence this time constant are of primary importance. Thus, sensor diameter, shown by eqn. 6.30 to be the dimension having greatest influence on the time constant, should be as small as possible. In addition, a time constant due to heat storage in the support needles becomes important unless the heat transfer to the support needles is a small fraction of the total. This implies that the aspect ratio should be as large as possible. Bremhorst and Gilmore (1978) claim that the aspect ratio for temperature sensors should be above 400.

Researchers show no preference for one sensor material over another for temperature probes. Bremhorst and Gilmore (1978) used wires of 90% platinum and 10% rhodium, and Bremhorst, Krebs, and Gilmore (1977) used platinum–iridium wire. Ming Ho, Jakus, and Parker (1976) used iridium wire for temperature measurements in flames.

Contamination of temperature sensors by dirt can be troublesome, and Larsen and Busch (1974) found temperature sensors used for atmospheric measurements to be much more likely to become fouled than similarly exposed velocity sensors operated at higher overheat ratios. The problem was found to be even more severe in a highly humid environment.

Not only is contamination more likely to occur on temperature sensors, but the layer of contaminants has an adverse effect on the frequency response. Smits, Perry, and Hoffman (1978) found that sensors used in temperature measurements have much lower rolloff frequency when contaminated. The sensor time constant was found to decrease markedly during use before finally reaching a constant value. Microscopic examination confirmed that this effect was due to the accumulation of small particles on the surface of the sensor. Figure 6.23 shows the decrease in sensor rolloff frequency with time due to contamination. Although this figure shows that the sensor reaches a stable rolloff frequency in three to four hours, wire sensors of very small diameter were found to require as long as 20 hours to stabilize. In some cases the sensor can be cleaned by using a higher sensor heating current for a short time to burn away the contaminants. Contamination of temperature sensors by heavy layers of salt were studied both experimentally and theoretically by Schacher and Fairall (1979).

## 6.7 Simultaneous measurement of velocity and temperature

If the velocity and temperature in a flow field vary independently, each can be measured separately. There are several methods that allow combined measurement of velocity and temperature by using one or more sensors, and all are based on common principles.

### The basic technique for combined measurements

Almost all of the techniques presented in this section are based on eqn. 3.21 for the velocity and temperature sensitivity of a heated sensor, and they require the use of a sensor at several overheat ratios, followed by a manipulation of the data to extract the velocity and temperature information (Corrsin, 1949). To understand how the velocity and temperature can be obtained in this way, eqn. 3.21 is squared and averaged to obtain

$$\overline{e_s^2} = S_{\text{vel}}^2 \overline{u^2} + 2 S_{\text{vel}} S_{\text{temp}} \overline{u t_f} + S_{\text{temp}}^2 \overline{t_f^2} \tag{6.30}$$

where $t_f$ is the fluctuating component of temperature.

If the velocity and temperature sensitivities have been experimentally determined beforehand, one sensor can be operated at three different overheat ratios (denoted in the following equations by subscripts 1, 2, and 3) to give three simultaneous equations similar to eqn. 6.30 – one for each overheat ratio. Because there are three unknowns, $\overline{u^2}$, $\overline{t_f^2}$, and $\overline{u t_f}$, enough equations are available to allow calculation of the unknowns at the completion of the test. The equations for velocity and temperature are

$$\overline{u^2} = \frac{A\overline{e_{s_1}^2} + B\overline{e_{s_2}^2} + C\overline{e_{s_3}^2}}{D}$$

and

$$\overline{t_f^2} = \frac{E\overline{e_{s_1}^2} + F\overline{e_{s_2}^2} + G\overline{e_{s_3}^2}}{D}$$

where

$$A = S_{temp2}S_{temp3}(S_{vel2}S_{temp3} - S_{vel3}S_{temp2})$$

$$B = S_{temp1}S_{temp3}(S_{vel3}S_{temp1} - S_{vel1}S_{temp3})$$

$$C = S_{temp1}S_{temp2}(S_{vel1}S_{temp2} - S_{vel2}S_{temp1})$$

$$E = S_{vel2}S_{vel3}(S_{vel2}S_{temp3} - S_{vel3}S_{temp2})$$

$$F = S_{vel1}S_{vel3}(S_{vel3}S_{temp1} - S_{vel1}S_{temp3})$$

$$G = S_{vel1}S_{vel2}(S_{vel1}S_{temp2} - S_{vel2}S_{temp1})$$

$$D = S_{temp1}^2(S_{vel2}^2 S_{vel3}S_{temp3} - S_{vel3}^2 S_{vel2}S_{temp2})$$
$$+ S_{temp2}^2(S_{vel3}^2 S_{vel1}S_{temp1} - S_{vel1}^2 S_{vel3}S_{temp3})$$
$$+ S_{temp3}^2(S_{vel1}^2 S_{vel2}S_{temp2} - S_{vel2}^2 S_{vel1}S_{temp1})$$

This method was found by Arya and Plate (1969) to allow a more accurate measurement of velocity than temperature. But because temperature is most accurately measured by using a low overheat ratio, one measurement can be taken this way. Equation 6.30 for this low overheat ratio is

$$\overline{e_{s1}^2} = S_{temp1}^2 \overline{t_f^2}$$

The other two overheat ratios lead to

$$\overline{e_{s2}^2} = S_{vel2}^2 \overline{u^2} + 2S_{vel2}S_{temp2}\overline{ut}_f + S_{temp2}\overline{t_f^2}$$

and

$$\overline{e_{s3}^2} = S_{vel3}^2 \overline{u^2} + 2S_{vel3}S_{temp3}\overline{ut}_f + S_{temp3}\overline{t_f^2}$$

which are solved simultaneously. An advantage of this method is the simplicity of the equations used. Then

$$\overline{u^2} = \frac{H\overline{e_{s1}^2} + I\overline{e_{s2}^2} + J\overline{e_{s3}^2}}{K}$$

and

$$\overline{t_f^2} = \frac{\overline{e_{s1}^2}}{S_{temp1}}$$

where

$$H = S_{temp2}S_{temp3}(S_{vel2}S_{temp3} - S_{vel3}S_{temp2})$$
$$I = S_{vel3}S_{temp1}^2 S_{temp3}$$
$$J = -S_{vel2}S_{temp1}^2 S_{temp2}$$

$$K = S_{vel2}S_{vel3}S_{temp1}^2(S_{vel2}S_{temp3} - S_{vel3}S_{temp2})$$

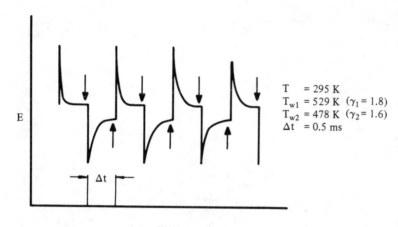

Figure 6.24. A graph of anemometer output voltage with the overheat ratio alternated between two values every 0.5 ms. Arrows indicate the point where data was taken. Reprinted with permission from U. Michel and E. Frobel, DFVLR Institute, Berlin.

### The use of one sensor at several overheat ratios

A simple way to make combined measurements of velocity and temperature is by operating a single sensor at several sequentially changed overheat ratios. This method is also less expensive, because only one anemometer is required.

Operation of the sensor at only three different overheat ratios was found to be inaccurate by Arya and Plate (1969); instead from six to eight different overheat ratios were used. This overdetermined the unknowns, giving several values of velocity and temperature. A least squares fit was used to obtain the final values.

It is possible to automatically vary the overheat ratio, rapidly cycling it over several values at a frequency above that of the variations in velocity and temperature occurring in the flow field. This idea was shown to be feasible by McConachie and Bullock (1975), who modeled the system with an analog computer. A constant temperature anemometer was modified by Fiedler (1978) to allow the overheat ratio to be switched alternately between two separate values every 0.5 ms. A representation of the change in bridge voltage with time during switching is shown in Figure 6.24. The bridge voltage is seen to reach a high value during the switching operation and then fall to a value representing operation at the new overheat ratio.

### Use of several sensors at different overheat ratios

If more than one sensor is used for combined measurement of velocity and temperature, the measurements can be made either simultaneously or more accurately.

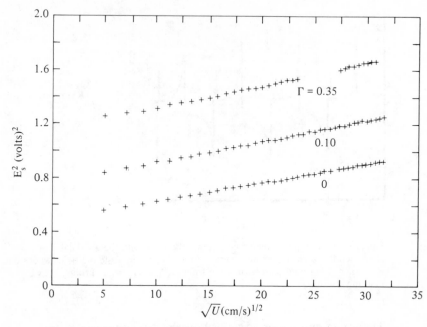

Figure 6.25. The heat loss characteristics for a hot wire sensor operated at low overheat ratio ($\Delta T = 50°C$) in a helium–air mixture at several values of gas mixture concentration, $\Gamma$. Reprinted with permission from J. Way and P. A. Libby, Hot-wire probes for measuring velocity and concentration in helium–air mixtures, *AIAA J.*, 8 (1970), 976–978.

The accuracy of these measurements can be improved if the spread of overheat ratios is large. This is done by operating one sensor as a temperature probe at low overheat ratio in the constant current mode. For higher overheat ratios, one probe can be used in the constant temperature mode and cycled between several values of overheat ratio (Arya and Plate, 1969).

Simultaneous measurement of velocity and temperature was obtained by Sakao (1973) using two sensors – one at an overheat ratio of 1.7, and the other at 1.2 – with an analog computer used to separate the temperature and velocity signals.

## 6.8 The measurement of gas mixture concentration

There are several ways to measure gas mixture concentration with a hot wire anemometer. A hybrid probe consisting of a film sensor in such close proximity to a wire sensor that thermal interference occurs can be used to measure both velocity and concentration. Or a single-sensor probe can be used to measure concentration if the fluid velocity is known. Finally, the gas mixture can be drawn through a choked orifice and across a wire sensor for a concentration measurement that is independent of fluid velocity.

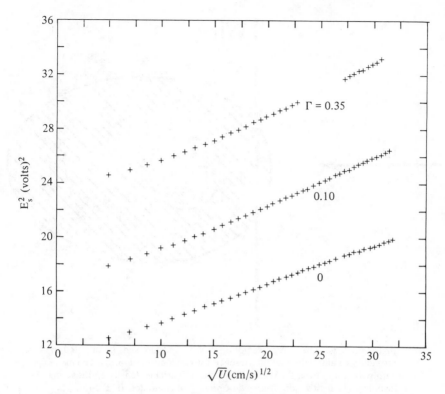

Figure 6.26. The heat loss characteristics for a hot film sensor operated at high overheat ratio ($\Delta T = 275°C$) in a helium–air mixture at several values of gas mixture concentration, $\Gamma$. Reprinted with permission from J. Way and P. A. Libby, Hot-wire probes for measuring velocity and concentration in helium–air mixtures, *AIAA J.*, 8 (1970), 976–978.

### The interfering sensor method

An analysis by Corrsin (1949) generated interest in the use of the hot wire anemometer to measure gas mixture concentration. It showed that a single sensor operated at several overheat ratios, or several sensors of different diameters, could be used for a combined measurement of velocity and gas mixture concentration.

Two cylindrical sensors of different diameters and at different overheat ratios were tested by Way and Libby (1970) in a helium–air mixture, and each sensor responded well to variations in gas mixture concentration, as shown in the next two figures. Figure 6.25 shows the heat loss characteristics for a wire sensor operated at a low overheat ratio at several helium–air mixture concentrations, and Figure 6.26 illustrates the heat loss characteristics for a cylindrical hot film sensor operated at a high overheat ratio. These figures show that since helium has higher thermal conductivity than air, for constant velocity an increase in helium concentration causes an increase in

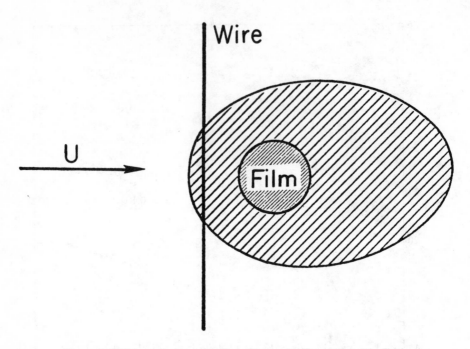

Figure 6.27. Schematic representation of wire and film sensors oriented in close proximity to allow thermal interference for concentration measurements. Reprinted with permission from P. A. Libby, Studies in variable-density and reacting turbulent shear flows, in *Studies in Convection* (Launder, B. E., Ed.), pp. 1–43, Academic Press, New York (1977).

the rate at which heat leaves the sensor and, thus, an increase in bridge voltage of the constant temperature anemometer.

If these two sensors are placed in the flow field at the same point, the King's law expression (eqn. 3.7) for each can be combined to eliminate velocity and give

$$E_w^2 = A(\Gamma) + B(\Gamma)E_f^2$$

where $\Gamma$ is the concentration of helium, $A$ and $B$ are constants, and $E_w$ and $E_f$ are the bridge voltage associated with the wire and film sensors, respectively. From this equation it can be seen that a graph of $E_w^2$ versus $E_f^2$ is a function of concentration only. This method lacks the sensitivity required to separate velocity and concentration, but Way and Libby improved sensitivity by moving the sensors closer together so that the thermal wake from the hotter sensor warmed the cooler sensor.

In this technique a hybrid X-array probe consisting of two sensors is used, each perpendicular to the velocity vector and to each other as well. The sensor arrangement is shown in Figure 6.27, where a hot wire sensor operating at a small overheat ratio is located upstream from a cylindrical hot

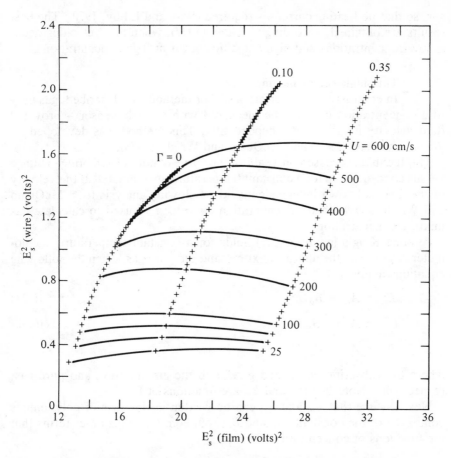

Figure 6.28. Calibration curve for thermally interfering sensors for concentration measurements in a helium–air mixture. Reprinted with permission from J. Way and P. A. Libby, Hot-wire probes for measuring velocity and concentration in helium–air mixtures, *AIAA J.*, 8 (1970), 976–978.

film sensor operating at a large overheat ratio. The two sensors are mutually perpendicular and close enough together to allow the downstream hot film sensor to contribute to the heating of the upstream hot wire sensor. Although the original analysis by Corrsin is not valid for this configuration, the resulting calibration curves, shown in Figure 6.28, indicate that both velocity and concentration can be found by measuring the output voltages from the two sensors.

Sensor spacing is somewhat dependent upon the velocities and concentrations encountered. For example, high velocity combined with low concentration cause the thermal field to be so close to the hot film sensor that thermal interference is not possible. Likewise, low velocity and high concentration cause the thermal field to expand, overheating the upstream sensor, so that no heating current is required (Way and Libby, 1970). The best results are obtained, according to Libby (1977), when low velocities occur at low concentrations and high velocities occur at high concentrations.

### The single-sensor method

In contrast to the interfering sensor method just described, gas mixture concentration can also be measured with a single sensor – provided fluid velocity is measured independently. This method was developed by Wasan and Baid (1971) and McQuaid and Wright (1973).

This technique is based on a calibration curve of normalized bridge voltage versus concentration for a constant temperature anemometer. If the velocity of the fluid is known independently, the following analysis from McQuaid and Wright shows that this calibration curve can be used to calculate gas mixture concentration.

Assume King's law (eqn. 3.7) holds for a constant temperature anemometer in pure air, the air–gas mixture, and the pure gas. Then the following equations apply:

$$E_a^2 = A_a + B_a U^n \tag{6.31}$$

$$E_m^2 = A_m + B_m U^n \tag{6.32}$$

$$E_g^2 = A_g + B_g U^n$$

where the subscripts $a$, $m$, and $g$ refer to the air, mixture, and pure gas, respectively. Note that $A_m$ and $B_m$ are functions of $\Gamma$.

Next assume that $A_m$ and $B_m$ can be written in the form of polynomials composed of the constants $A_a$ and $B_a$ from eqn. 6.31 plus other terms that are functions of concentration. Thus

$$A_m = A_a + a_1\Gamma + a_2\Gamma^2 + a_3\Gamma^3 + \cdots \tag{6.33}$$

$$B_m = B_a + b_1\Gamma + b_2\Gamma^2 + b_3\Gamma^3 + \cdots \tag{6.34}$$

where $a_1$, $a_2$, $a_3$, . . . and $b_1$, $b_2$, $b_3$, . . . are constants. Next assume the bridge voltage, fluid velocity, and concentration to be composed of both a mean and a fluctuating part, and $E = \overline{E} + e$, $U = \overline{U} + u$, and $\Gamma = \overline{\Gamma} + \gamma$. Combine eqn. 6.32 with eqns. 6.33 and 6.34 and neglect higher-order terms.

$$\begin{aligned}
\overline{E}_m^2 + 2e\overline{E}_m = {}& [A_a + a_1(\overline{\Gamma} + \gamma) + a_2(\overline{\Gamma}^2 + 2\gamma\overline{\Gamma}) + \cdots] \\
& + [B_a + b_1(\overline{\Gamma} + \gamma) + b_2(\overline{\Gamma}^2 + 2\gamma\overline{\Gamma}) + \cdots]\overline{U}^n \\
& + [B_a + b_1(\overline{\Gamma} + \gamma) + b_2(\overline{\Gamma}^2 + 2\gamma\overline{\Gamma}) + \cdots]nu\overline{U}^{n-1}
\end{aligned} \tag{6.35}$$

Equation 6.32 for mean values only is

$$\overline{E}_m^2 = \overline{A}_m + \overline{B}_m \overline{U}^n \tag{6.36}$$

where

$$\overline{A}_m = A_a + a_1\overline{\Gamma} + a_2\overline{\Gamma}^2 + \cdots$$

$$\overline{B}_m = B_a + b_1\overline{\Gamma} + b_2\overline{\Gamma}^2 + \cdots$$

Subtract eqns. 6.35 and 6.36 to get

$$2e\overline{E}_m = n\overline{B}_m\overline{U}^n\frac{u}{\overline{U}}$$

$$+ [(a_1\overline{\Gamma} + 2a_2\overline{\Gamma}^2 + \cdots) + (b_1\overline{\Gamma} + 2b_2\overline{\Gamma}^2 + \cdots)\overline{U}^n]\frac{\gamma}{\overline{\Gamma}}$$

which can be written as

$$e = S_{\text{vel}}\frac{u}{\overline{U}} + S_{\text{conc}}\frac{\gamma}{\overline{\Gamma}}$$

where

$$S_{\text{vel}} = \frac{n\overline{B}_m\overline{U}^n}{2\overline{E}_m} \tag{6.37}$$

$$S_{\text{conc}} = \frac{(a_1\overline{\Gamma} + 2a_2\overline{\Gamma}^2 + \cdots) + (b_1\overline{\Gamma} + 2b_2\overline{\Gamma}^2 + \cdots)\overline{U}^n}{2\overline{E}_m} \tag{6.38}$$

Substitute eqn. 6.32 into eqns. 6.37 and 6.38 to get

$$S_{\text{vel}} = \frac{n(\overline{E}_m^2 - A_m)}{2\overline{E}_m}$$

and

$$S_{\text{conc}} = \frac{p\overline{\Gamma} + q\overline{\Gamma}^2 + r\overline{\Gamma}^3 + \cdots}{2\overline{E}_m}$$

where

$$p = a_1 + b_1\overline{U}^n, \quad q = 2(a_2 + b_2\overline{U}^n), \quad r = 3(a_3 + b_3\overline{U}^n)$$

Write eqn. 6.31 using mean quantities

$$\overline{E}_a^2 = A_a + B_a\overline{U}^n$$

Form the difference

$$\overline{E}_m^2 - \overline{E}_a^2 = (a_1\overline{\Gamma} + a_2\overline{\Gamma}^2 + a_3\overline{\Gamma}^3 + \cdots)$$

$$+ (b_1\overline{\Gamma} + b_2\overline{\Gamma}^2 + b_3\overline{\Gamma}^3 + \cdots)\overline{U}^n$$

which can be rearranged to give

$$\overline{E}_m^2 - \overline{E}_a^2 = p\overline{\Gamma} + \tfrac{1}{2}q\overline{\Gamma}^2 + \tfrac{1}{3}r\overline{\Gamma}^3 + \cdots$$

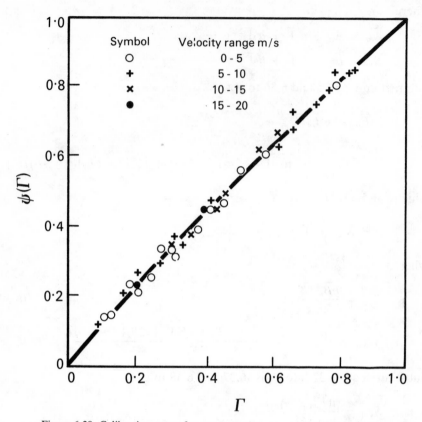

Figure 6.29. Calibration curve for concentration measurements in an argon–air mixture. Reprinted with permission from J. McQuaid and W. Wright, The response of a hot-wire anemometer in flows of gas mixtures, *Int. J. Heat Mass Trans.*, 16 (1973), 819–828.

The normalized quantity

$$\psi = \frac{\overline{E_m^2} - \overline{E_a^2}}{\overline{E_g^2} - \overline{E_a^2}}$$

can be shown experimentally to be virtually independent of velocity. It is written as

$$\psi = \left(\frac{p}{\overline{E_g^2} - \overline{E_a^2}}\right)\overline{\Gamma} + \frac{1}{2}\left(\frac{q}{\overline{E_g^2} - \overline{E_a^2}}\right)\overline{\Gamma}^2 + \cdots$$

plotted versus concentration, as shown in Figure 6.29, and used as a calibration curve for gas mixture concentration measurements.

An extension of this method has been suggested by McQuaid and Wright to allow combined measurements of the fluctuating components of velocity and concentration in some gas mixtures by using two sensors operated side by side.

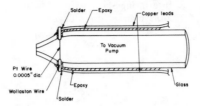

Figure 6.30. Gas mixture concentration probe consisting of a hot wire sensor downstream from a sonic orifice through which the gas mixture is drawn. Reprinted with permission from G. L. Brown and M. R. Rebollo, A small, fast-response probe to measure composition of a binary gas mixture, *AIAA J.*, 10 (1972), 649 – 652.

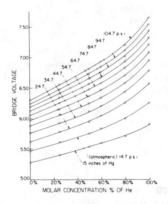

Figure 6.31. Calibration curve for the gas mixture concentration sensor shown in Figure 6.30 for helium in nitrogen. Reprinted with permission from G. L. Brown and M. R. Rebollo, A small, fast-response probe to measure composition of a binary gas mixture, *AIAA J.*, 10 (1972), 649–652.

**The sonic nozzle and sensor method**

Gas mixture concentration can be measured without knowing fluid velocity by drawing a sample of the fluid from the flow field with a small nozzle attached to a suction hose (Brown and Rebollo, 1972). The nozzle, illustrated in Figure 6.30, contains a small orifice at its tip through which the gas mixture is drawn. The gas reaches sonic velocity at the orifice and then losses speed as it passes through the larger-diameter section of the tube before reaching the hot wire sensor. The sensor operates in the constant temperature mode, and the output voltage is a function of both the pressure inside the tube and the gas mixture concentration.

A calibration curve for this nozzle in a helium–air mixture is shown in Figure 6.31. A similar technique, using a capillary tube instead of an orifice, was developed by Adler (1972).

The time constant for a sonic orifice and sensor modeled as a half-Rankine body with inside diameter $d$, an orifice at its tip, and small suction was found

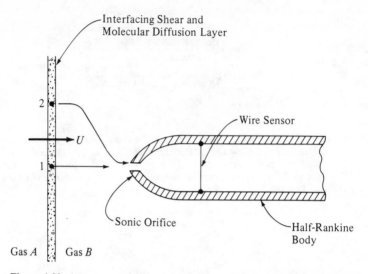

Figure 6.32. A concentration probe is shown sampling the interface between two gases. Reprinted with permission from A. E. Perry, The time response of an aspirating probe in gas sampling, *J. Phys. E.: Sci. Instr.*, 10 (1977), 898–902.

by Perry (1977) to be

$$\tau = 0.075 \, \frac{d}{U}$$

where $U$ is the free stream velocity.

To see that a time constant exists, assume two pure gases that contact each other and mix along a vertical line are swept along in the horizontal direction toward the nozzle, as shown in Figure 6.32. When suction is applied, pure gas B is first drawn into the orifice, followed by the mixture located at point 1. The mixture at point 2, however, must travel a greater distance and enters the orifice at a slightly later time.

## 6.9 Compressible flow measurements

Very different problems are encountered when the hot wire anemometer is used in compressible flow. Not only does hydrodynamic loading tend to deflect the probe support and damage the sensor, but some of the concepts considered standard in hot wire anemometry no longer apply.

### Heat transfer in compressible flow

When an unheated sensor is placed in a compressible flow, aerodynamic heating causes the sensor temperature to rise from ambient to a value called the equilibrium temperature (sometimes referred to as the recovery temperature) $T_e$. If the sensor is then heated electrically, its temperature rises to its heated temperature, $T_s$. The equilibrium temperature is easy to measure and is often used in several nondimensionalized forms. One, called the temperature loading, is a ratio of the sensor temperature and the equilibrium temperature, $T_e$, defined as

$$\text{Temperature loading} = \frac{T_s - T_e}{T_e}$$

A second, the temperature ratio, is a ratio of the equilibrium temperature and the stagnation temperature, defined as

$$\text{Temperature ratio} = \frac{T_e}{T_o}$$

When the sensor is placed in the flow, a detached bow shock forms upstream. This shock wave is normal to the streamlines that pass in the vicinity of the sensor, so it measures the characteristics of the subsonic flow field directly behind a normal shock wave. The Reynolds number behind a normal shock wave is therefore an important parameter and is defined as

$$\text{Re}_2 = \frac{\rho_2 U_2 d}{\mu_2}$$

where $d$ is the diameter of the sensor, and the subscript 2 refers to conditions downstream from a normal shock wave.

A graph of the temperature ratio versus the Reynolds number behind the shock, shown in Figure 6.33, is almost completely independent of Mach number. The graph can also be separated by a vertical line at Re $\approx$ 20 into two distinct regimes. For $\text{Re}_2 > 20$ the equilibrium temperature is almost constant and virtually independent of Reynolds number. For $\text{Re}_2 < 20$ the equilibrium temperature increases sharply as the Reynolds number decreases, rising to a value greater than the stagnation temperature at low Reynolds numbers (Lauffer and McClellan, 1956).

The Nusselt number for a sensor in compressible flow is a function of the following quantities (Kovasznay, 1950 and Lauffer and McClellan, 1956):

$$\text{Nu} = \text{Nu}\left(\text{Re}_\infty, M_\infty, \text{Pr}_\infty, k_f, \frac{T_s - T_e}{T_e}, \frac{l}{d}\right)$$

where the subscript $\infty$ refers to free stream conditions. If the Prandtl number and specific heat ratio can be shown to be constant and the aspect ratio can be corrected for heat conduction to the support needles by using one of the methods discussed previously, this equation reduces to

$$\text{Nu} = \text{Nu}\left(\text{Re}_\infty, M_\infty, \frac{T_s - T_e}{T_e}\right) \tag{6.39}$$

There is some disagreement about whether the Nusselt number should be evaluated at the stagnation temperature or the temperature after the normal

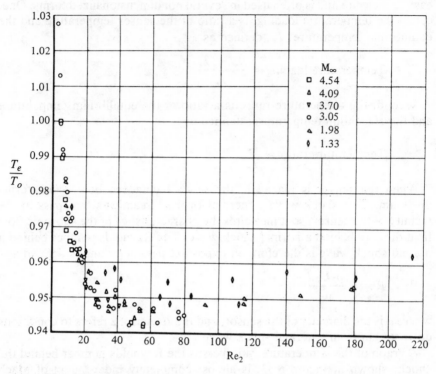

Figure 6.33. The variation in temperature ratio for compressible flow. Reprinted with permission from J. Laufer and R. McClellan, Measurement of heat transfer from fine wires in supersonic flows, *J. Fluid Mech.*, 1 (1956), 276–289.

shock, but if the latter is chosen, eqn. 6.39 becomes

$$\mathrm{Nu}_2 = \mathrm{Nu}\left(\mathrm{Re}_2, M_2, \frac{T_s - T_e}{T_e}\right)$$

The Nusselt number for the sensor in compressible flow is evaluated by modifying eqn. 3.12 to give

$$\mathrm{Nu}_2 = \frac{\alpha I^2 R_s R_e}{\pi l k_f (R_s - R_e)[1 + \alpha(T_e - T_{\mathrm{ref}})]}$$

where $T_{\mathrm{ref}} = 273°\mathrm{K}$, and $R_e$ is the resistance of the sensor at the equilibrium temperature. The thermal conductivity, $k_f$, should be evaluated at the temperature downstream from the normal shock.

Laufer and McClellan (1956) found the Nusselt number downstream from the shock to be a function of only the Reynolds number downstream from the shock and the temperature loading or

$$\mathrm{Nu}_2 = \mathrm{Nu}\left(\mathrm{Re}_2, \frac{T_s - T_e}{T_e}\right)$$

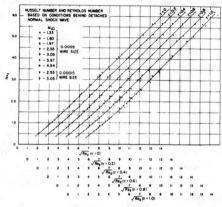

Figure 6.34. The Nusselt number variation at several temperature loadings for compressible flow. Reprinted with permission from J. Laufer and R. McClellan, Measurement of heat transfer from fine wires in supersonic flows, *J. Fluid Mech.,* 1 (1956), 276–289.

as shown in Figure 6.34, where the Nusselt number is plotted versus the square root of the Reynolds number.

### Mean velocity measurements

The general procedure for making mean velocity measurements in compressible flow requires that static pressure, $P$, stagnation pressure, $P_o$, and stagnation temperature, $T_o$, first be measured by some means other than hot wire anemometry. Then several calibration curves are constructed and used to calculate mass flow, Mach number, mean velocity, and density from the experimental data.

This procedure, developed by Laufer and McClellan (1956), begins by selecting a preaged probe that has been tested to assure that its sensor will not stretch while in use. This probe is placed in an oven, and its resistance versus temperature characteristics are graphed. Next, with the probe in the free stream, make a graph of $T_e/T_o$ versus the equilibrium Reynolds number, $Re_e$, defined as

$$Re_e = Re_2 \left( \frac{\mu_2}{\mu_e} \right)$$

where $\mu_2$ and $\mu_e$ represent the absolute viscosity measured at the temperature after the shock and the equilibrium temperature, respectively. To do this, the Mach number is held constant, and the Reynolds number is varied to graph $T_e/T_o$ versus $Re_2$. Repeating this for other values of Mach numbers shows all data to fall approximately on the same curve, indicating that $T_e/T_o$ is a function of $Re_2$ only and is independent of the free stream Mach number.

Next, the sensor is heated to temperature $T_s$, and the Nusselt number,

$\mathrm{Nu}_s$, is calculated for the sensor. This is done by evaluating the thermal conductivity at the temperature of the heated sensor. End-loss corrections are applied to $\mathrm{Nu}_s$, and a graph similar to Figure 6.34 is made, except that the sensor Nusselt number is plotted versus the square root of the sensor Reynolds number for a variety of temperature loadings. The sensor Reynolds number is defined as

$$\mathrm{Re}_s = \mathrm{Re}_2 \left( \frac{\mu_2}{\mu_s} \right)$$

where $\mu_s$ is the absolute viscosity measured at the sensor temperature. The finished graph shows the sensor Nusselt number to be a function only of Reynolds number and temperature loading and, thus, is independent of the free stream Mach number.

After these two graphs are constructed by using the sensor in the free stream, measurements of mass flow, Mach number, mean velocity, and density are taken in the boundary layer of the test model. The sensor is positioned in the flow, and $T_e$ is measured. The sensor is then heated to an operating temperature chosen beforehand, and $\mathrm{Nu}_s$ is calculated. This value of $\mathrm{Nu}_s$ is located on the graph of $\mathrm{Nu}_s$ versus $\sqrt{\mathrm{Re}_s}$, and the value of $\mathrm{Re}_s$ is read. From this, the mass flow, $\rho U$, is calculated. Next, $\mathrm{Re}_e$ is calculated, and the graph of $T_e/T_o$ versus $\mathrm{Re}_e$ is used to find $T_o$. Then, because static pressure can be assumed to remain constant across the boundary layer being measured, the Mach number, mean velocity, and density can be measured.

### Fluctuation measurements

Measurement of velocity, temperature, and density fluctuations are made in compressible flow by using methods developed by Kovasznay (1950, 1953) and Morkovin (1956). These measurements are based on the use of a single sensor operated at several overheat ratios. In the following analysis the notation and format follow in part that of Kistler (1959).

If the fluctuating voltage signal from a constant temperature or constant current hot wire anemometer is assumed to be a linear combination of the mass flow and the stagnation temperature, then

$$\Delta E = S_{\rho U} \frac{\Delta (\rho U)}{\overline{\rho U}} + S_{T_o} \frac{\Delta T_o}{\overline{T_o}} \tag{6.40}$$

where $S_{\rho U}$ is the mass flow sensitivity, and $S_{T_o}$ is the stagnation temperature sensitivity, and the new symbology for mean and fluctuating bridge voltage, mass flow, and stagnation temperature are $E = \overline{E} + \Delta E$, $\rho U = \overline{\rho U} + \Delta (\rho U)$, and $T_o = \overline{T_o} + \Delta T_o$, respectively. These sensitivities are found experimentally for each sensor. Square eqn. 6.40, average, and divide by the square of the stagnation temperature sensitivity to get

$$\frac{\overline{\Delta E^2}}{S_{T_o}} = \left( \frac{S_{\rho U}}{S_{T_o}} \right)^2 \frac{\overline{\Delta (\rho U)^2}}{(\rho U)^2} + 2 \left( \frac{S_{\rho U}}{S_{T_o}} \right) \frac{\overline{\Delta (\rho U) \Delta T_o}}{\rho U T_o} + \frac{\overline{\Delta T_o^2}}{T_o^2} \tag{6.41}$$

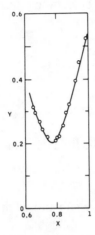

Figure 6.35. A typical fluctuation diagram with notation following Morkovin (1956). Reprinted with permission from A. J. Laderman and A. Demetriades, Hot-wire measurements of hypersonic boundary-layer turbulence, *Phys. Fluids*, 16 (1973), 179–181.

The general equation for a hyperbola is $x^2 = Ay^2 + By + C$, where $A$, $B$, and $C$ are constants. Note that eqn. 6.41 is the equation for a hyperbola.

The three unknowns in eqn. 6.41 are $\Delta(\rho U)^2/(\rho U)^2$, $\Delta T_o^2/T_o^2$, and $\Delta(\rho U)\Delta T_o/\rho U T_o$, and if the sensor is operated at at least three different overheat ratios, then, because the sensitivities are different for each overheat ratio, the three resulting equations can be solved simultaneously. Better results are obtained if more than three overheat ratios are used. Laderman and Demetriades (1973, 1974), for example, used a minimum of 15 different overheat ratios in their experiments.

A least squares method allows the best fit of a hyperbola through the data and completely defines eqn. 6.41. A typical graph, called a fluctuation diagram, is shown in Figure 6.35.

This procedure can be used to calculate the mass flow and the stagnation temperature, but further analysis is required to find the velocity, static temperature, and density of the flow.

If the density and velocity are separated into mean and fluctuating quantities such as $\rho = \overline{\rho} + \Delta\rho$ and $U = \overline{U} + \Delta U$, then

$$\Delta(\rho U) = \overline{\rho}\Delta U + \overline{U}\Delta\rho \tag{6.42}$$

Dividing by the mean values, squaring, and averaging gives

$$\frac{\overline{\Delta(\rho U)^2}}{(\overline{\rho U})^2} = \frac{\overline{\Delta U^2}}{\overline{U}^2} + \frac{\overline{\Delta\rho^2}}{\overline{\rho}^2} + 2\frac{\overline{\Delta\rho\Delta U}}{\overline{\rho U}}$$

This is the first of the equations used to find the desired quantities.

Next, the stagnation temperature relationship of eqn. 6.41 is written in terms of the static temperature in the following way: The relationship between stagnation and static temperature for compressible flow (Shapiro, 1953, p. 83) is

$$\frac{T_o}{T} = 1 + \left(\frac{\gamma - 1}{2}\right) M^2$$

where

$$M^2 = \frac{U^2}{\gamma R T}$$

and $R$ is the gas constant for the fluid. Combining these two equations gives

$$T_o = T + \frac{\gamma - 1}{2\gamma R} U^2 \tag{6.43}$$

Mean and fluctuating values are substituted and mean values subtracted to get

$$\Delta T_o = \Delta T + \frac{\gamma - 1}{\gamma R} \overline{U} \Delta U \tag{6.44}$$

Eqn. 6.43 for mean values is divided into eqn. 6.44 and then squared and averaged to obtain

$$\frac{\overline{\Delta T_o^2}}{\overline{T_o^2}} = \left\{\frac{1}{1 + [(\gamma - 1)/2]M^2}\right\}^2 \frac{\overline{\Delta T^2}}{\overline{T}^2}$$

$$+ \left\{\frac{(\gamma - 1)M^2}{1 + [(\gamma - 1)/2]M^2}\right\}^2 \frac{\overline{\Delta U^2}}{\overline{U}^2}$$

$$+ \frac{2(\gamma - 1)M^2}{\{1 + [(\gamma - 1)/2]M^2\}^2} \frac{\overline{\Delta T \Delta U}}{TU}$$

This is the second equation to be used later.

From eqns. 6.42 and 6.44 we get

$$\Delta(\rho U) \Delta T_o = (\rho \Delta U + \overline{U} \Delta \rho) \left[\Delta T + \left(\frac{\gamma - 1}{\gamma R}\right) \overline{U} \Delta U\right]$$

Dividing by mean values gives

$$\frac{\overline{\Delta(\rho U)} \overline{\Delta T_o}}{\rho U T_o} = \left\{\frac{1}{1 + [(\gamma - 1)/2]M^2}\right\} \frac{\overline{\Delta T \Delta U}}{TU}$$

$$+ \left\{\frac{(\gamma - 1)M^2}{1 + [(\gamma - 1)/2]M^2}\right\} \frac{\overline{\Delta U^2}}{\overline{U}^2}$$

$$+ \left\{\frac{1}{1 + [(\gamma - 1)/2]M^2}\right\} \frac{\overline{\Delta \rho \Delta T}}{\rho T}$$

$$+ \left\{\frac{(\gamma - 1)M^2}{1 + [(\gamma - 1)/2]M^2}\right\} \frac{\overline{\Delta \rho \Delta U}}{\rho U}$$

This is the third equation needed.

But these equations contain four unknowns – too many to be solved simultaneously. One more equation is needed, and often the "no sound" assumption (that is, for the flow in question the pressure fluctuations are small compared to the velocity, static temperature, and density fluctuations) can be used. This means

$$\frac{\Delta P}{\overline{P}} \ll \frac{\Delta U}{\overline{U}}, \frac{\Delta \rho}{\overline{\rho}}, \frac{\Delta T}{\overline{T}}$$

To use this assumption, mean and fluctuating values are substituted into the perfect gas law:

$$\frac{\Delta P}{\overline{P}} = \frac{\Delta \rho}{\overline{\rho}} + \frac{\Delta T}{\overline{T}}$$

Under the assumption of negligible pressure fluctuations, this equation becomes

$$\frac{\Delta \rho}{\overline{\rho}} = -\frac{\Delta T}{\overline{T}}$$

which is the fourth and last equation needed to find velocity, static temperature, and density. A method of correcting the fluctuation measurements for heat loss from the sensor to the support needles has been developed by Ko, McLaughlin, and Troutt (1978).

### Shock tube measurements

When a probe is used in a shock tube, the support needles remain at ambient temperature because supersonic flow occurs only for a fraction of a second; at other times the support needles are in a quiescent gas. By contrast, the support needles of a probe in a supersonic wind tunnel are at the recovery temperature. Because of this difference, a probe cannot be calibrated in a supersonic wind tunnel and then used with accuracy in a shock tube.

Despite these support needle temperature problems, there are several ways that a hot wire sensor can be used in a shock tube. A common approach is to mount uncalibrated probes at equally spaced intervals along the tube to measure the time of flight of the shock wave. Or several uncalibrated sensors can be located along the diameter of the shock tube to measure shock from curvature (Gude and Christoffersen, 1968).

Another technique, developed by Guy (1971), utilizes the fact that after the diaphragm in a shock tube is punctured, a compression wave moves away from the diaphragm in one direction and a rarefaction wave having the same mass flux, $\rho U$, but different temperature moves away in the other direction. A probe positioned on one side of the diaphragm will measure the mass flux in the compression wave but at an elevated temperature. Repositioning the same probe on the other side of the diaphragm and repeating the experiment allows a measurement in the expansion wave flow of the same mass flux at a reduced temperature. The correct mass flux can be calculated from these two measurements.

### Probe design for compressible flow

The high velocities experienced by a probe used in compressible flows cause problems that are seldom encountered in other types of hot wire measurements. Most commonly encountered are sensor stretching and breakage caused by hydrodynamic loading and failure due to high temperatures.

Sensor breakage is often caused by the hydrodynamic forces on the sensor, but breakage by vibration or the impact of airborne dirt particles should not be ruled out. A wire sensor can also stretch because of the hydrodynamic loading, and this can permanently alter the cold resistance and invalidate the calibration curve. For hypersonic flow measurements Dewey (1961) discarded test data taken with hot wire sensors having a change in cold resistance of only a few tenths of 1% measured before and after each test run. Laufer and McClellan (1956) did the same for supersonic flow measurements.

Even if a permanent deformation of the sensor does not occur, transient changes in its length can cause errors. This "strain-gauging" effect was noted by Spangenberg (1955) using platinum and platinum–rhodium hot wire sensors having diameters from 1.3 μm to 7.6 μm. Breakage and permanent deformation were also a problem.

Sensor breakage or permanent deformation is more likely to occur with the sensor mounted taut between the support needles. Instead, according to Spangenberg (1955), the sensor should have a slight sag along its length. Spangenberg used a sag from 0.2 mm to 0.3 mm for a sensor length of 3 mm for hot wire sensors placed normal to the mean velocity vector, and found inclined wires to need no sag. For hot wire sensors about 1 mm long, a barely perceptible sag was found to perform well.

Although sensors having a high aspect ratio are more likely to break or deform in use, a high aspect ratio is necessary to maintain a more constant temperature distribution along the sensor and reduce the percentage of heat lost to the support needles. Dewey (1961) achieved best results in hypersonic flow with platinum–iridium wire sensors having an aspect ratio range of $320 < l/d < 360$.

Most cold-resistance drift problems can be solved by annealing the wires before use. Dewey (1961) annealed 2.5-μm-diameter wires of 90% platinum and 10% iridium by heating the sensor electrically to a dull glow for a few minutes. Spangenberg (1955) annealed wires of the same material and diameter in a two-step process. First, the wires were preheated electrically to 2.05 times their cold resistance. Then the sensor current was gradually increased over the next three to five minutes to raise the sensor resistance to about 2.4 times cold resistance. This was found to eliminate cold-resistance drift for the life of the probe. Spangenberg (1955) also strength tested new batches of wire by clamping samples in a fixture and subjecting them to maximum wind loading while at operating temperature. Wire batches that failed this test were discarded.

In addition to failure by breaking or stretching, the sensor can also overheat and melt; or the solder holding it to the support needles can melt. In the latter case the sensor can be welded to the support needles. Spangenberg (1955) found heat loading to be so great that heat loss from the 3.8-mm-diameter platinum wires to the support needles caused the solder connections to melt. The wires were subsequently arc-welded in place to eliminate this problem.

In hypersonic wind-tunnel flows at $M > 5$, the stagnation temperature is often maintained as high as $T_o \approx 540°C$. This places a severe restriction on probe design, even when the sensor is unheated. For some measurements several different overheat ratios are used; this means that sensor temperatures must be chosen to maximize the difference between each overheat ratio without using a sensor temperature that is so high the probe will be damaged.

At high temperatures a cooled sensor probe can be used. This type of probe was used in hypersonic flow by Ellington and Trottier (1968), with both Silicon Oil 704 and Flourolube FS used as coolant fluids with sensor temperatures as high as 800°K. They recommend use of high-viscosity coolants to maintain laminar flow through the sensor, because transition in the coolant from laminar to turbulent flow was found to increase anemometer output voltage noise by one order of magnitude. Water, when used as a coolant, was found to boil at sensor temperatures in excess of 500°K. The temperature distribution on the cooled sensor is skewed, being lower at the end of the sensor where the coolant enters.

Often, a streamlined probe body is designed for use in a supersonic flow. An example is the commercially manufactured probe for compressible flow measurements shown in Figure 6.36.

An unusual hot film probe was developed by Roos and Bogar (1982) to measure the position and movement of a shock wave. The probe is shown in Figure 6.37, where a glass cylinder having a pointed end has a spiral film sensor deposited along its length. The anemometer output voltage is a fairly linear function of the location of a shock wave on the cylinder, as shown by the calibration curve of Figure 6.38.

Gottesdiener (1980) was able to eliminate the need for end-loss corrections by using a probe having heated support needles. A thermistor and thermocouple were glued to the ends of the 0.02-mm-diameter, 20-mm-long, hot wire sensor, and the ends of the sensor were held at the temperature of the center of the sensor. The support needles were stainless steel tubes and contained the lead from the sensor wire and two wires each from the thermistor and thermocouple.

When an X-array probe is used in a compressible flow, the relationship between the location of the sensor and that of the Mach waves in the flow

Figure 6.36. A commercially manufactured hot wire probe for compressible flow measurements having a wedge-shaped probe body. Reprinted with permission from TSI, Inc.

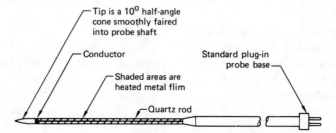

Figure 6.37. A hot film probe used to measure shock wave location and movement. The sensor is a spiral metal film deposited around the quartz cylinder. Reprinted with permission from F. W. Roos and T. J. Bogar, Comparison of hot-film probe and optical techniques for sensing shock motion, *AIAA J.*, 20 (1982), 1071–1076.

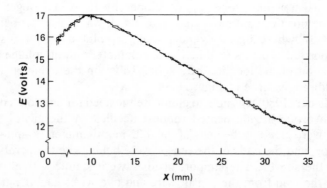

Figure 6.38. A calibration curve for the shock wave location probe shown in Figure 6.37. In this graph, $x$ is the distance along the probe. Reprinted with permission from F. W. Roos and T. J. Bogar, Comparison of hot-film probe and optical techniques for sensing shock motion, *AIAA J.*, 20 (1982), 1071–1076.

must be considered. The sensor should not lie along a Mach wave (Kovasznay, 1950), and at $M > 1.4$ the Mach waves are angled at 45° to the mean velocity vector. Because it is standard practice in turbulence measurements to locate an X-array probe with its sensors at ±45° to the mean velocity vector, it is possible for a sensor to lie along a Mach wave attached to the tip of a support needle (Bradshaw, 1971, p. 160).

## 6.10 Two-phase flow measurements

There are several types of two-phase flow measurements that are possible with a hot wire anemometer. The hot wire sensor can be used to measure the collision frequency of liquid droplets in a gas, and the droplet size can be inferred as well. Void fraction and the frequency at which bubbles of gas in a liquid strike a sensor can also be measured. Finally, the hot wire sensor can be used to measure the velocity of a gas close to a liquid surface where waves occasionally cover the sensor.

### Liquid-droplet measurements in air

When a liquid droplet in air strikes the sensor of a hot wire or hot film probe, the sensor is suddenly cooled. Then the sensor heats the liquid droplet, and it finally evaporates. The variation in anemometer output voltage during impact can be observed, and the frequency at which droplets strike the sensor can be measured. This may lead to a determination of their size as well. Liquids used for droplets in these experiments have been water, safflower oil (Goldschmidt and Eskinazi, 1966), Sinco-Prime 70 (Goldschmidt and Householder, 1969), dibutyl-phthalate (Bragg and Tevaarwerk, 1974), and ice crystals, methanol, and ethylene glycol (Vonnegut and Neubauer, 1952).

A detailed study of the sensor–droplet interaction was made by Bragg and Tevaarwerk (1974) by illuminating the sensor with a stroboscope and viewing the droplet impact with a microscope. After impact, the droplet clings to the sensor and gets smaller by evaporation and the action of the air stream. In addition, because the temperature profile is not uniform along the sensor, droplets striking near the sensor ends take longer to evaporate than particles of the same size that strike near the midpoint. The anemometer output voltage signal during impact shows a rapid and almost linear rise to a maximum value, followed by a slower decline. The droplet signature is sometimes a function of the liquid used.

Goldschmidt and Eskinazi (1966) found the turbulent eddy sizes in their experiments to be larger than the droplets and were thus able to use a bandpass filter to eliminate the turbulence signals. Each droplet signal was digitized to generate a square pulse for each droplet strike as an aid in counting the rate at which droplets strike the sensor.

As explained by Delhaye (1969), not all particles moving toward the sensor cause a measurable output voltage change. Instead, some move toward the

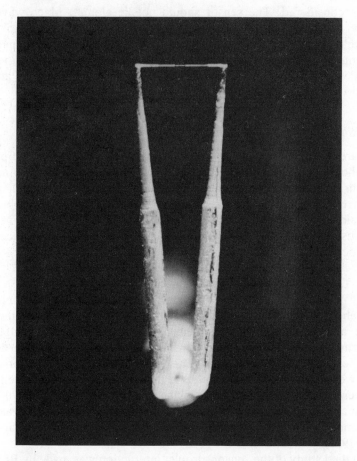

Figure 6.39. A hot wire probe for void fraction measurements in freon. The closeup view of the sensor shows that only a small section of the middle of the wire is unplated. Reprinted with permission from G. E. Dix, Vapor void fractions for forced convection with subcooled boiling at low flow rates, General Electric Co., NEDO-10491, 1971.

sensor and then follow the flow around the sensor without striking. Other droplets strike the sensor but cause output voltage changes that are too small to be detected. Goldschmidt and Eskinazi (1966) define an "impact coefficient" to describe this phenomenon as the ratio of the number of particles colliding with the sensor to the number of particles occurring in an area equal to the frontal area of the sensor.

Although there is no question that droplet impact frequency can be measured with a hot wire anemometer, there is no agreement that droplet size can be inferred from the output voltage signal. Bragg and Tevaarwerk (1974) could see no correlation between droplet size and output voltage signal, and Goldschmidt and Householder (1969) found that for droplets smaller in di-

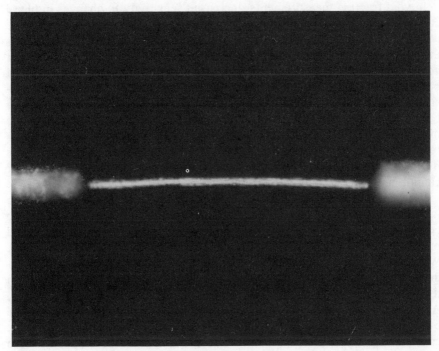

Figure 6.39. (*cont.*)

ameter than 200 μm at moderate velocities, the peak voltage during impact was linearly proportional to the diameter of the liquid droplet.

### Void fraction measurements

The extent of bubbles in a fluid can be measured by allowing bubbles carried along by the fluid to strike a hot wire or hot film sensor. Since the thermal conductivity for a liquid is usually much greater than for gas, less heating current is required if a gas bubble encloses the sensor, and the output voltage will indicate a sharp dip as the bubble collides. The fraction of the time bubbles enclose the sensor is a measure of the local void fraction.

Both hot wire and hot film probes have been used in void fraction measurements. Delhaye (1969) favored the commercial hot film probe because of its resistance to fouling and reliability. Shiralkar (1970), however, use an uninsulated tungsten hot wire probe in freon, which is an electrical insulator, and Herringe and Davis (1974) found hot wire probes to give faster response than hot film probes but recommended the use of larger-diameter sensors than are commonly used in air measurements in order to prevent breakage. Dix (1971) used a tungsten hot wire sensor that was 3.8 μm in diameter and 0.127 mm long in freon. This sensor was modified by plating the ends to allow only a small fraction of its length to be sensitive to velocity. Figure 6.39 is a view of this probe along with a closeup view of the sensor.

In void fraction measurements, increased sensitivity to changes in phase can be obtained if the sensor is operated at a temperature that causes nucleate boiling on its surface. This deviation from common practice is advantageous in void fraction measurements because of the great difference in heat transfer coefficients between nucleate boiling and vapor flow (Delhaye, 1969). If boiling is not wanted, the overheat ratio can be adjusted for maximum sensitivity to the bubbles. Dix (1971) found "bubble noise" caused by nucleate boiling to be a problem but reduced it to an acceptable level by proper choice of overheat ratio. Bubbles formed by nucleate boiling were also found to remain attached to the sensor at low flow rates instead of being swept away by the fluid. Another disadvantage of nucleate boiling is that the sensor cannot be used for simultaneous measurement of liquid velocity.

When a bubble moves vertically upward through a quiescent liquid, the interaction of the bubble with a stationary sensor actually begins before contact occurs. As the bubble approaches the sensor, the liquid just ahead of it is pushed forward, and the anemometer output voltage begins to rise. When the bubble reaches the sensor, puncture does not take place immediately. Delhaye (1969) found the tip of conical hot film sensors to deform the bubble upon contact, pushing it in extensively before breaking through. As the probe tip moved through the interior of the bubble and neared the other surface, that surface was seen to dimple inward to meet the probe tip, causing the probe to pierce this surface before the bubble traveled one bubble diameter.

Bremhorst and Gilmore (1976a), who used a hot wire probe for void fraction measurements, classified several types of sensor–bubble interactions. In the ideal case, a "direct hit," the entire length of the sensor is enclosed by the bubble, and the sensor passes through the bubble center. A "glancing hit" occurs when the sensor is completely enclosed by the bubble, but the sensor does not pass through the center of the bubble. The bubble is said to make a "partial hit" when only part of the sensor enters the bubble. Another phenomenon noted by Bremhorst and Gilmore was an attractive force between the bubble and the sensor that caused the bubble to remain attached to the sensor after passing, stretching the bubble, and forming a "detachment tail" that was noticeable only at bubble speeds below about 1 m/s.

Sometimes the bubble is cut in half by the hot wire sensor. When this happens, the sensor does not enter the bubble, but, instead, the bubble is pressed against the sensor and deforms until one bubble surface is pushed against the surface at the other side of the bubble. When the two surfaces touch, the bubble separates into two smaller bubbles. Sometimes, after a bubble is cut in two, they recombine a short distance from the sensor (Bremhorst and Gilmore, 1976a). A stereo microphotograph of bubbles hitting a hot wire probe is shown in Figure 6.40.

Toral (1981) filmed the bubble–sensor interaction at 2000 frames/s for air or ethanol vapor bubbles in liquid ethanol. An inspection of the photographs

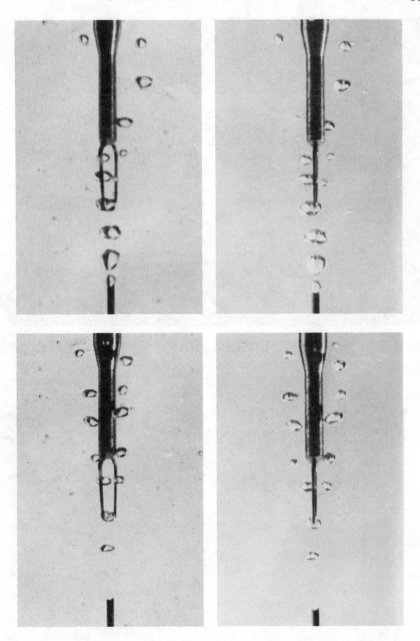

Figure 6.40. Stereo microphotographs of two hot wire probes positioned in a bubbly flow to measure local void fraction. Reprinted with permission from K. Bremhorst and D. B. Gilmore, Response of hot wire anemometer probes to a stream of air bubbles in a water flow, *J. Phys. E.: Sci. Instr.*, 9 (1976), 347–357.

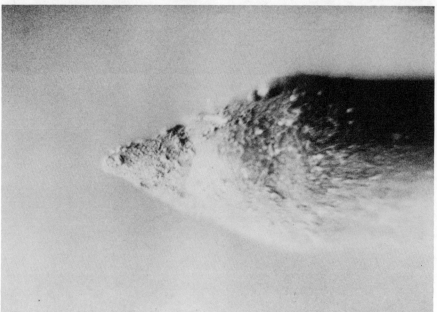

Figure 6.41. A conical hot film probe is shown in unused condition in the upper photograph and after exposure to insufficiently filtered mercury–nitrogen flow in the lower photograph. Reprinted with permission from P. Gherson and P. S. Lykoudis, Hot film anemometry in a two-phase (liquid metal–gas) medium, in *Seventh Biennial Symp. Turb.* (Zakin, J. L. and Patterson, G. K., Eds.), University of Missouri, Rolla.

showed bubbles to decrease in speed by about 33% as they passed over the sensor. There was also a distortion in bubble shape during impact. The combination of these two effects was found to give an underestimation of about 9% in bubble velocity.

Two-phase flow measurements in mercury containing bubbles of nitrogen were made by Gherson and Likoudis (1981). Fouling was, of course, a major problem that was aggravated by the gas bubbles that increased the free surface area where contaminants collect. In addition, fouling material on the sensor was found to trap small bubbles of gas to give a further reduction in sensitivity. A comparison between clean and fouled conical hot film probes is shown in Figure 6.41. Cylindrical hot film probes were tried for these measurements, but small gas bubbles were held in the wake behind the sensor, and this influenced the measurements.

### Measurements in air near the water surface

It is possible to make velocity measurements in air with the sensor located so close to the water surface that waves occasionally cover the sensor (Wills, 1976). When this happens, the rate of sensor cooling increases because water has a higher coefficient of thermal conductivity than air. A constant temperature anemometer compensates for this sudden cooling by increasing the heating current to maintain the sensor at the same temperature, and the output voltage rises suddenly as the sensor is immersed. While underwater the sensor may be covered by air bubbles carried along as the sensor passed through the surface of the water, and it may not be able to measure water velocity. When the probe enters the air again, water will remain on the sensor until it evaporates. This evaporation was found by Wills to require about 0.004 s, after which the air velocity could again be measured.

An unusual type of sensor burnout can occur if the sensor is partially submerged with one end in the water and the other in the air. Then the heating current increases to compensate for increased cooling of the submerged part of the sensor, causing the part in the air to burn out.

# 7 WAKE-SENSING ANEMOMETERS

In this chapter we look at techniques for measuring the magnitude and direction of the velocity vector by sensing the wake convected downstream from a heated sensor. In the first method, a heated sensor is used to measure fluid speed in the usual way, and a temperature sensor is positioned downstream to sense the direction of the thermal wake and, thus, the direction of the mean velocity vector. A second method uses an intermittently heated wire to generate a hot spot in the fluid that is convected past a temperature sensor. The time of flight is a function of the fluid velocity.

## 7.1 Sensing the heated wake direction

An X-array probe is the usual choice when a measurement of the angle of the mean velocity vector is needed. But the X-array probe requires two electronics packages and a complicated data-analysis procedure in order to determine speed and direction. An alternative method is to use a hot wire sensor with two temperature sensors downstream to form a triangular array of mutually parallel sensors, as shown in Figure 7.1. With this configuration the angle at which the heated wake leaves the sensor can be found.

The advantage of this array, according to Walker and Bullock (1972), becomes evident if a measurement is needed near a wall. Here, the X-array probe is more likely to disturb the flow and suffer from inaccuracies caused by velocity gradients along the length of its sensors.

Disadvantages of wake-sensing probes include errors due to wake distortion by fine-scale turbulence and the possibility that eddy shedding could invalidate the data. Also, there is a transport delay time equal to the time required for the fluid to travel from the heated sensor to the cold sensor. Also, for certain sensor orientations, buoyancy effects could cause errors, but Rey and Beguier (1977) found that this was not a problem at velocities above 0.2 m/s.

A variation of this wake-sensing technique is to use a thermocouple to sense the heated wake. Such a device, developed by Sawatzki (1971), is illustrated in Figure 7.2. In this design the distance between the sensor and the thermocouple can be varied by a micrometer-head adjustment – a feature that might be too bulky if implemented on a smaller probe.

A typical wake-sensing probe of the three-wire type has a heated wire located about 20 diameters upstream from two cold wires, which are smaller

188

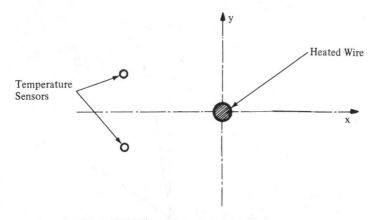

Figure 7.1. A sensor configuration for measurement of pitch angle. The wake of the larger-diameter heated wire is sensed by the temperature wires downstream. Reprinted with permission from C. Rey and C. Beguier, On the use of a three parallel wire probe, *DISA Info.*, 21 (1977), 11–15.

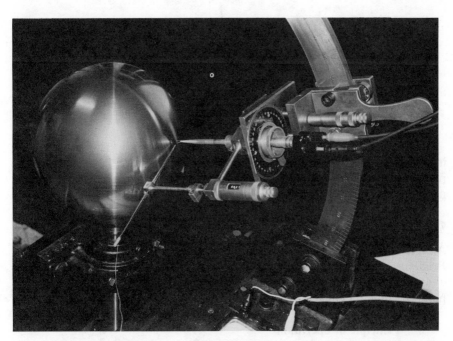

Figure 7.2. A standard hot wire probe with an adjustable holder used to position a thermocouple downstream for simultaneous measurements of the speed and direction of the flow in the vicinity of a sphere. Reprinted with permission from O. Sawatzki, Measurements in the disturbed laminar boundary layer of a special three-dimensional flow field, *DISA Info.*, 11 (1971), 5–24.

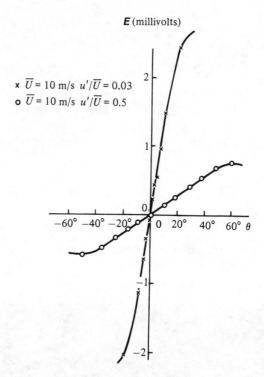

Figure 7.3. The calibration curve of a wake-sensing probe. Reprinted with permission from C. Rey and C. Beguier, On the use of a three parallel wire probe, *DISA Info.*, 21 (1977), 11–15.

in diameter. A typical calibration curve, shown in Figure 7.3, illustrates the response of a wake-sensing probe of this type to changes in the angle of the mean velocity vector.

## 7.2 The pulsed wire anemometer

An anemometer operating on the time-of-flight principle was developed by Bradbury and Castro (1971). It is commercially produced and promises to overcome some of the limitations of the constant current and constant temperature anemometers.

### Principle of operation

The pulsed wire anemometer operates by heating a portion of the fluid under investigation and timing the travel of the "temperature spot" as it is convected downstream past a temperature sensor. To implement this technique, a wire is heated electrically by either a sine wave or a series of discrete pulses. A second hot wire sensor, operated at low overheat ratio and located a short distance downstream, is used to detect the temperature spots as they are convected past. The time between the generation of the temperature spots and their arrival at the second wire is inversely proportional to the fluid velocity.

The pulsed wire anemometer has some intriguing advantages. For example, the time of flight of the temperature spot is not influenced by the ambient temperature of the fluid, so, unlike hot wire anemometers of other types, no temperature compensation scheme is required. There is a linear relationship between the time of flight of the temperature spot and the fluid velocity, which means no electronic linearizers are needed. Although the sensors of a pulsed wire anemometer are just as susceptible to contamination as the sensors of other types of hot wire anemometers, the fouled pulsed wire probe causes no loss in accuracy until catastrophic fouling causes the system to stop functioning altogether. Also, because the pulsed wire probe is usually designed with a second temperature sensor upstream from the heated wire, the forward–reverse ambiguity can be resolved. Finally, pulsed wire anemometers can be used to measure mean velocity, turbulence intensity, and probability distribution.

The most obvious limitation of the pulsed wire anemometer is its inability to make continuous velocity measurements. Instead, it takes discrete measurements in time, as does the laser Doppler velocimeter. A problem common to all pulsed wire methods is heat storage in the heated wire, which limits the pulse frequency to about 50 Hz. Another limitation is electrical coupling between the heated wire and the cold wires, which causes the heating current signal to be sensed by the cold wire before it senses the temperature spot.

Several types of pulsed wire anemometers have been reported in the literature. The most common type has its wire heated by discrete current pulses. This instrument was described by Bauer (1965): Two standard hot wire probes were used – one to heat the fluid, and the second positioned downstream to sense the temperature spots. This instrument was developed, and its potential was analyzed by Bradbury and Castro (1971).

A pulsed wire anemometer using a sinusoidal heating current was developed by Kovasznay (1949). Again, two separate standard hot wire probes were used, and an oscilloscope was used to observe Lissajous patterns made from the sinusoidal heating current and the heated wake detected downstream. This allowed measurement of phase lag and, hence, fluid velocity. Walker and Westenberg (1956) modified the test procedure by moving the temperature sensor downstream until a straight line was displayed as a Lissajous figure, indicating the two signals to be either exactly in phase or out of phase by 180°. This position was noted, and the temperature sensor was moved farther downstream until a straight line was again observed. The fluid velocity was then found from the equation

$$U = 2\pi f \frac{\Delta x}{\Delta \phi}$$

where $f$ is the frequency of the heating current sine wave, $\Delta x$ is the distance between the two sensors, and $\Delta\phi$ is the change in phase angle noted at the two positions of the sensors. Walker and Westenberg obtained best results with both sensors perpendicular to the mean velocity vector and to each other as well.

A third type of pulsed wire anemometer, developed by Kielbasa, Rysz, Smolarski, and Stasicki (1972), uses a two-wire probe, with one wire used to heat the fluid and the other used to sense the temperature spots. But in this case, electronic circuitry controls the wire heating current, so that it is high when the temperature at the downstream sensor is low, and vice versa. This causes the system to oscillate; the frequency of the heating current pulses is directly proportional to the fluid velocity.

In the rest of this chapter the type of anemometer with a discretely pulsed heating current is considered.

### Probes for pulsed wire anemometry

The typical pulsed wire anemometer probe having a discretely pulsed wire is shown in Figure 2.21. Notice that two temperature sensors are used – one upstream, and the other downstream and perpendicular to the heated wire. This arrangement is less sensitive to variations in the width of the heated wake, a consideration of importance for measurements in turbulent flow (Bradbury and Castro, 1971). In practice, this probe is placed in the flow field with all wires normal to the mean velocity vector.

The wires of a pulsed wire probe are generally of the same material and have the same diameter as other types of hot wire probes, except for the center wire, which is often larger in diameter so that it can be heated to a high temperature with less chance of breaking. For example, in the probe design of Bradbury and Castro (1971), the center wire was 10 μm in diameter and made either of platinum or nickel, and the temperature-sensing wires were 5 μm in diameter and made of platinum; wire lengths were about 6.25 mm.

The spacing between the wires depends upon the velocity expected. Bradbury and Castro (1971) used a spacing of 1.25 mm between the center wire and the flanking wires, which limited the maximum speed to about 14 m/s.

A novel characteristic of the pulsed wire anemometer is the audible "click" heard each time a heating current pulse is generated. This has the advantage of reassuring the user that the center wire and pulsing circuit are not defective.

Thermal and mechanical stresses in the pulsed wire can cause the entire probe to vibrate and the center wire to fail. Tombach (1973) was able to extend probe life considerably by cementing a ruby watch jewel (it has the shape of a small ring) to the tip of each support needle. The center wire passes through the hole in the center of the jewel and is soldered to the side of the support needle, allowing the wire to flex during use. Alternatively,

Bradbury and Castro (1971) placed a slight kink in the center wire to allow enough slack for conventional attachment.

Because small, closely spaced, high-temperature spots are required, the wire should be pulsed at high frequency and short duration. But if the pulse frequency is too high, the wire will not cool sufficiently between pulses and burn out. In practice, this limits the pulse rate to about 50 pulses/s, as reported by Bauer (1965) and Gaster and Bradbury (1976). Pulse durations of about 1 μs or 2 μs and a heating current of several amps are sometimes used.

An inevitable effect associated with the pulsed wire probe is electromagnetic coupling between the pulsed wire and the temperature sensors, which give signals in the millivolt range on the temperature sensor when the center wire is pulsed. This is far greater than the temperature signal, which has a magnitude of about 100 μV (Bradbury and Castro, 1971). At high fluid velocities this voltage spike can interfere with the temperature signal.

### The calibration of pulsed wire probes

The calibration curve for the pulsed wire anemometer is linear, or almost so, and can be expressed by the following calibration equation, used by Bauer (1965):

$$U = \frac{h}{T}$$

$h$ is the distance between the heated wire and the temperature sensor, and $T$ is the time of flight. Bradbury (1976) found the calibration curve to deviate slightly from linear and used least squares to fit the following equation to the data:

$$U = \frac{A}{T} + \frac{B}{T^2}$$

where $A$ and $B$ are constants. A typical calibration curve is shown in Figure 7.4. Two different calibration curves are obtained, one for each direction of the flow, because the two temperature sensors are not equidistant from the heated wire.

As a temperature spot approaches the temperature sensor, its output voltage initially rises nonlinearly, and this is followed by a linear increase that tails off to a maximum value. But at what instant can the temperature spot be considered to have reached the temperature sensor? If the signal is processed by a Schmitt-trigger circuit producing a square wave when the temperature signal reaches a predetermined level, uniform results can be obtained. Bauer (1965) used a back extrapolation scheme on the oscilloscope trace of the temperature sensor output voltage to calculate temperature spot arrival without a Schmitt trigger. The sloping linear portion of the trace displayed on the screen of a storage oscilloscope was continued back with

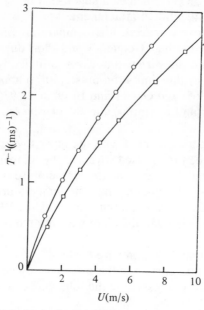

Figure 7.4. A typical pulsed wire anemometer calibration curve for both flow directions. Reprinted with permission from L. J. S. Bradbury, Measurements with a pulsed-wire and a hot-wire anemometer in the highly turbulent wake of a normal flat plate, *J. Fluid Mech.*, 77 (1976), 473–497.

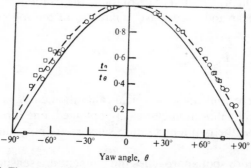

Figure 7.5. The yaw response of the pulsed wire anemometer probe. The solid line is a cosine curve, and the dotted line represents eqn. 7.1. with $\epsilon = 0.1$. Reprinted with permission from L. J. S. Bradbury. Measurements with a pulsed-wire and a hot-wire anemometer in the highly turbulent wake of a normal flat plate, *J. Fluid Mech.*, 77 (1976), 473–497.

a straight line until the horizontal axis was intersected. The point of inter-section was considered by Bauer to be the instant at which the temperature spot reached the temperature sensor.

It takes longer for the temperature spot to reach the sensor with the probe placed at an angle to the flow, and this causes sensitivity to angular variations. As shown in Figure 7.5, this angular response is almost a cosine function. Bradbury (1976) found the response of the time of flight to the yaw angle to be

$$\frac{t_0}{t_\theta} = \cos \theta + \epsilon \sin \theta \tag{7.1}$$

where $t_0$ is the mean time of flight at zero yaw angle, $t_\theta$ is the mean time of flight in yaw, $\theta$ is the yaw angle, and $\epsilon \approx 0.1$. This equation was found to apply for mean velocities in the range of 0.25 m/s $< U <$ 15 m/s.

Castro and Cheun (1982) used the following yaw-response equation, which is equivalent to eqn. 7.1:

$$U_{\text{eff}} = U(\cos \theta + k \sin \theta)$$

where $k$ is a yaw factor equal to about 0.1.

# REFERENCES

Adler, D. (1972). A hot wire technique for continuous measurement in unsteady concentration fields of binary gaseous mixtures. *J. Phys. E.: Sci. Instr.*, 5, 163–169.

Aggarwal, J. K. (1974). Development of a hot film gage suitable for the measurement of a three-dimensional velocity vector. *J. Phys. E.: Sci. Instr.*, 7, 733–737.

(1978). A directionally insensitive thin-film shear-stress gauge. *J. Phys. E.: Sci. Instr.*, 11, 349–352.

Almquist, P., and Legath, E. (1965). The hot-wire anemometer at low air velocities. *DISA Info.*, 2, 3–4.

Andreas, E. L. (1979). Analysis of crossed hot-film velocity data. *DISA Info.*, 24, 15–23.

Anhalt, J. (1973) Device for in-water calibration of hot-wire and hot-film probes. *DISA Info.*, 15, 25–26.

Antonia, R. A., Browne, L. W. B., and Chambers, A. J. (1981). Determination of time constants of cold wires. *Rev. Sci. Instr.*, 52, 1382–1385.

Antonini, G., Guiffant, G., and Perrot, P. (1977). The use of holography for direct visualization of thermal distribution around hot-wires in a liquid under dynamic conditions. *DISA Info.*, 21, 28–32.

Arya, S. P. S., and Plate, E. J. (1969). Hot-wire measurements in non-isothermal flow. *Inst. and Cont. Syst.*, 42, 87–90.

Astarita, G., and Nicodemo, L. (1969). Behavior of velocity probes in viscoelastic dilute polymer solutions. *Ind. Engr. Chem. Fund.*, 8, 582–585.

Aydin, M., and Leutheusser, H. J. (1980). Very low velocity calibration and application of hot-wire probes. *DISA Info.*, 25, 17–18.

Bauer, A. B. (1965). Direct measurement of velocity by hot-wire anemometry. *AIAA J.*, 3, 1189–1191.

Bearman, P. W. (1971). Corrections for the effect of ambient temperature drift on hot-wire measurements in incompressible flow. *DISA Info.*, 11, 25–30.

Bellhouse, B. J., and Bellhouse, F. H. (1968). Thin-film gauges for the measurement of velocity or skin friction in air, water or blood. *J. Phys. E.: Sci. Inst.*, 1, 1211–1213.

Bellhouse, B. J., and Rasmussen, C. G. (1968). Low-frequency characteristics of hot-film anemometers. *DISA Info.*, 6, 3–10.

Bellhouse, B. J. and Schultz, D. L. (1966). Determination of mean and dynamic skin friction separation and transition in low speed flow with a thin film heated element. *J. Fluid Mech.*, 24, 379–400.

Bellhouse, B. J., Schultz, D. L., and Karatzas, N. B. (1966). The measurement of fluctuating components of velocity and skin friction with thin-film heated elements, with application in water, air and blood flows. University of Oxford, Dept. Engr. Sci. Rept. No. 1003.

196

Bertrand, J., and Couders, J. P. (1978). Hot-film probe calibration in liquids. *DISA Info.*, 23, 28–32.

Betchov, R., and Welling, W. (1952). Some experiences regarding the nonlinearity of hot wires. NACA TM 1223.

Blackshear, P. L., and Fingerson, L. M. (1962). Rapid-response heat flux probe for high temperature gases. *ARS J.*, 32, 1709–1715.

Blackwelder, R. F., and Kaplan, R. E. (1976). On the wall structure of the turbulent boundary layer. *J. Fluid Mech.*, 76, 89–112.

Blick, E. F., Sabbah, H. N., and Stein, P. D. (1975). Red blood cells and turbulence. In *Turbulence in Liquids* (G. K. Patterson, and J. L. Zakin, Eds.), pp. 17–21. University of Missouri, Rolla.

Bond, A. D. and Porter, A. M. (1967). Self-aligning hot-wire probe. *J. Roy. Aero. Soc.*, 71, 657–658.

Bonis, M., and van Thinh, N. (1973). A heat transfer law for a conical hot-film probe in water. *DISA Info.*, 14, 11–14.

Bradbury, L. J. S. (1976). Measurements with a pulsed-wire and hot-wire anemometer in the highly turbulent wake of a normal flat plate. *J. Fluid Mech.*, 77, 473–497.

Bradbury, L. J. S., and Castro, I. P. (1971). A pulsed-wire technique for measurements in highly turbulent flows. *J. Fluid Mech.*, 49, 657–691.

Bradshaw, P. (1971). *An Introduction to Turbulence and Its Measurement.* Pergamon Press, Oxford.

Bragg, G. M., and Tevaarwerk, J. (1974). The effect of a liquid droplet on a hot wire anemometer probe. In *Flow, Its Measurement and Control in Science and Industry*, Vol. 1, pp. 599–603. Instrument Society of America, Pittsburgh.

Bremhorst, K., and Gilmore, D. B. (1976). Comparison of dynamic and static hot wire anemometer calibrations for velocity perturbation measurements. *J. Phys. E.: Sci. Inst.*, 9, 1097–1100.

(1976a). Response of hot wire anemometer probes to a stream of air bubbles in a water flow. *J. Phys. E.: Sci. Inst.*, 9, 347–357.

(1978). Influence of end conduction on the sensitivity to stream temperature fluctuations of a hot-wire anemometer. *Int. J. Heat Mass Trans.*, 21, 145–154.

Bremhorst, K., Krebs, L., and Gilmore, D. B. (1977). The frequency response of hot-wire anemometer sensors to heating current fluctuations. *Int. J. Heat Mass Trans.*, 20, 315–322.

Brodowicz, K., and Kierkus, W. T. (1966). Experimental investigation of laminar free-convection flow in air above horizontal wire with constant heat flux. *Int. J. Heat Mass Trans.*, 9, 81–93.

Brown, G. L. and Rebollo, M. R. (1972). A small, fast-response probe to measure composition of a binary gas mixture. *AIAA J.*, 10, 649–652.

Bruun, H. H. (1971). Interpretation of a hot wire signal using a universal calibration law. *J. Sci. Instr.*, 4, 225–231.

(1976). A note on static and dynamic calibration of constant temperature hot-wire probes. *J. Fluid Mech.*, 76, 145–155.

(1978). Multi-probes and higher moments. In *Proc. Dyn. Flow Conf.*, pp. 943–961. Skovlunde, Denmark.

Buchhave, P. (1978). Transducer techniques. In *Proc. Dyn. Flow Conf.*, pp. 427–463. Skovlunde, Denmark.

Castro, I. P., and Cheun, B. S. (1982). The measurement of Reynolds stresses with a pulsed-wire anemometer. *J. Fluid Mech.*, 118, 41–58.

Champagne, F. H., and Sleicher, C. A. (1967). Turbulence measurements with inclined hot-wires. Part 2: Hot-wire response equations. *J. Fluid Mech.*, 28, 177–182.

Champagne, F. H., Sleicher, C. A., and Wehrmann, O. H. (1967). Turbulence measurements with inclined hot-wires. Part 1: Heat transfer experiments with inclined hot-wire. *J. Fluid Mech.*, 28, 153–176.

Chang, P. C., Grove, A., Achley, B., and Plate, E. J. (1970). A self-adjusting probe positioner for measuring flow fields in the vicinity of wind generated surface waves. *Rev. Sci. Instr.*, 41, 1544–1549.

Christensen, O. (1970). New trends in hot-film probe manufacturing. *DISA Info.*, 9, 30–36.

Cole, J., and Roshko, A. (1954). Heat transfer from wires at Reynolds numbers in the Oseen range. *Proc. Heat Trans. Fluid Mech. Inst.*, University of California, Berkeley.

Collis, D. C., and Williams, M. J. (1959). Two-dimensional convection from heated wires at low Reynolds numbers. *J. Fluid Mech.*, 6, 357–384.

Comte-Bellot, G. (1975). The physical background for hot-film anemometry. In *Turbulence in Liquids* (G. K. Patterson, and J. L. Zakin, Eds.), pp. 1–13. University of Missouri, Rolla.

(1976). Hot-wire anemometry. *An. Rev. Fluid Mech.*, 8, 209–231.

(1977). Hot wire and hot film anemometers. In *Measurement of Unsteady Fluid Dynamic Phenomena* (B. E. Richards, Ed.) pp. 123–162. Hemisphere, Washington.

Comte-Bellot, G., Strohl, A., and Alcaraz, E. (1971). On aerodynamic disturbances caused by single hot-wire probes. *J. Appl. Mech.*, 38, 767–774.

Corrsin, S. (1949). Extended applications of the hot-wire anemometer. NACA TN 1864.

(1963). Turbulence: Experimental methods. In *Handbuch der Physik*, Vol. 8, pp. 524–590.

Dahm, M., and Rasmussen, C. G. (1969). Effect of wire mounting system on hot-wire probe characteristics. *DISA Info.*, 7, 19–24.

Davies, P. O. A. L., and Fisher, M. J. (1964). Heat transfer from electrically heated cylinders. *Proc. Roy. Soc. A*, 280, 486–527.

Davies, P. O. A. L., Tanner, P. L., and Day. D. (1968). The Mark I linearised hot wire anemometer. University of Southampton, ISHV Report No. 159.

Davis, M. R. (1970). The dynamic response of constant resistance anemometers. *J. Phys. E.: Sci. Inst.*, 3, 15–20.

Delhaye, J. M. (1969). Hot-film anemometry in two-phase flow. In *Two-Phase Flow Instrumentation*, pp. 58–69. American Society of Mechanical Engineers, New York.

Delleur, J. W., Toebes, G. H., and Liu, C. L. (1968). Turbulence measurements in liquids: Some considerations and experiences regarding optimization. In *Advances in Hot Wire Anemometry* (W. L. Melnik, and J. R. Weske, Eds.), pp. 153–166. University of Maryland, College Park.

Dewey, C. F. (1961). Hot wire measurements in low Reynolds number hypersonic flows. *ARS J.*, 31, 1709–1718.

DISA Elektronik A/S probe catalog, 1980.

Dix, G. E. (1971). Vapor void fractions for forced convection with subcooled boiling at low flow rates. General Electric Co., NEDO-10491.

Doughman, E. L. (1972). Development of a hot-wire anemometer for hypersonic turbulent flows. *Rev. Sci. Instr.*, 43, 1200–1202.

Dring, R. P., and Gebhart, B. (1969). Hot-wire anemometer calibration for measurements at very low velocity. *J. Heat Trans.*, 91, 241–244.

Drubka, R. E., Tan-atichat, J., and Nagib, H. M. (1977). Analysis of temperature compensating circuits for hot wires and hot films. *DISA Info.*, 22, 5–14.

Dryden, H. L. (1936). Air flow in the boundary layer near a plate. NACA TR 562.

Dryden, H. L., and Kuethe, A. M. (1929). The measurement of fluctuations of air speed by the hot-wire anemometer. NACA Report 320.

Eckelmann, H. (1972). Hot-wire and hot-film measurements in oil. *DISA Info.*, 13, 16–22.

Eckelmann, H., Nychas, S. G., Brodkey, R. S., and Wallace, J. M. (1977). Vorticity and turbulence production in pattern recognized turbulent flow structures. *Phys. Fluids*, 20, S225–S230.

Eklund, T. I., and Dobbins, R. A. (1977). Application of the hot wire anemometer to temperature measurement in transient gas flows. *Int. J. Heat Mass Trans.*, 20, 1051–1057.

Ellington, E., and Trottier, G. (1968). Some observations on the application of cooled-film anemometry to the study of the turbulent characteristics of hypersonic wakes. In *Advances in Hot Wire Anemometry* (W. L. Melnik, and J. R. Weske, Eds.), pp. 52–64. University of Maryland, College Park.

Elsner, J., and Gundlach, W. R. (1973). Some remarks on the thermal equilibrium equation of hot-wire probes. *DISA Info.*, 14, 21–24.

Fabris, G. (1978). Probe and method for simultaneous measurements of "true" instantaneous temperature and three velocity components in turbulent flow. *Rev. Sci. Instr.*, 49, 654–664.

Fabula, A. G. (1968). Operating characteristics of some hot film velocity sensors in water. In *Advances in Hot Wire Anemometry* (W. L. Melnik, and J. R. Weske, Eds.), pp. 167–193. University of Maryland, College Park.

Fairall, C. W., and Schacher, G. (1977). Frequency response of hot wires used for atmospheric turbulence measurements in the marine environment. *Rev. Sci. Instr.*, 48, 12–17.

Fiedler, H. (1978). On data acquisition in heated turbulent flows. In *Proc. Dyn. Flow Conf.*, pp. 81–100. Skovlunde, Denmark.

Forman, K. M. (1969). Design and integration of turbulence experiment for mesocaphe 'Ben Franklin.' ASME paper 69-WA/UnT-9.

(1971). Some operational characteristics of hot film water velocity sensors. In *Proc. Symp. on Flow*, pp. 623–629. Pittsburgh, Paper 2-2-53.

Foss, J. F. (1978). Transverse vorticity measurements. In *Proc. Dyn. Flow Conf.*, pp. 983–1001. Skovlunde, Denmark.

(1981). Advanced techniques for transverse vorticity measurements. In *Seventh Biennial Symp. Turb.* (J. L. Zakin, and G. K. Patterson, Eds.), University of Missouri, Rolla.

Fox, J., Webb, W. H., Jones, B. G., and Hammitt, A. G. (1967). Hot-wire measurements of wake turbulence in a ballistic range. *AIAA J.*, 5, 99–102.

Francis, M. S., Kennedy, D. A., and Butler, G. A. (1978). Technique for the measurement of spatial vorticity distributions. *Rev. Sci. Instr.*, 49, 617–623.

Freymuth, P. (1967). Feedback control theory for constant-temperature hot-wire anemometers. *Rev. Sci. Instr.*, 38, 677–681.

Friehe, C. A., and Schwarz, W. H. (1968). Deviations from the cosine law for yawed cylindrical anemometer sensors. *J. Appl. Mech.*, 90, 655–662.

(1969). The use of pitot-static tubes and hot-film anemometers in dilute polymer solutions. In *Proc. Symp. Viscous Drag Reduction* (C . S. Wells, Ed.), pp. 281–296. Plenum, New York.

Fugita, H., and Kovasznay, L. S. G. (1968). Measurement of Reynolds stress by a single rotated hot wire anemometer. *Rev. Sci. Instr.*, 39, 1351–1355.

Gallagher, B. (1973). Calibration of probes in water at velocities under one meter per second. *DISA Info.*, 14, 19–20.

Gardner, R., and Lykoudis, P. (1971). Magnetofluidmechanic pipe flow in a transverse magnetic field. Part 1: Isothermal flow. *J. Fluid Mech.*, 47, 737–764.

Gaster, M., and Bradbury, L. J. S. (1976). The measurement of spectra of highly turbulent flows by a randomly triggered pulsed-wire anemometer. *J. Fluid Mech.*, 77, 499–509.

Geremia, J. O. (1972). Experiments on the calibration of flush mounted film sensors. *DISA Info.*, 13, 5–10.

Gessner, F. B., and Moller, G. L. (1971). Response behavior of hot wires in shear flow. *J. Fluid Mech.*, 47, 449–468.

Gherson, P., and Lykoudis, P. S. (1981). Hot film anemometry in a two-phase (liquid metal-gas) medium. In *Seventh Biennial Symp. Turb.* (J. L. Zakin and G. K. Patterson, Eds.), University of Missouri, Rolla.

Gilmore, D. C. (1967). The probe interference effect of hot wire anemometers. McGill University Mechanical Engineering Research Laboratory, TN 67-3.

Giovanangeli, J. P. (1980). A nondimensional heat transfer law for a slanted hot-film in water flow. *DISA Info.*, 25, 6–9.

Goldschmidt, V. W., and Eskinazi, S. (1966). Two-phase turbulent flow in a plane jet. *J. Appl. Mech.*, 33, 735–747.

Goldschmidt, V. W., and Householder, M. K. (1969). The hot wire anemometer as an aerosol droplet size sampler. *Atmos. Envir.*, 3, 643–651.

Gottesdiener, L. (1980). Hot wire anemometry in rarefied gas flow. *J. Phys. E. : Sci. Inst.*, 13, 908–913.

Grant, H. L., Stewart, R. W., and Moilliet, A. (1962). Turbulence spectra from a tidal channel. *J. Fluid Mech.*, 12, 241–263.

Grosh, R. J., and Cess, R. D. (1958). Heat transfer to fluids with low Prandtl numbers for flow across plates and cylinders of various cross sections. *Trans. ASME*, 80, 667–676.

Gude, K. E., and Christoffersen, J. A. (1968). The shock front curvature in a shock tube measured with hot-wire anemometers. *DISA Info.*, 6, 11–23.

Guitton, D. E., and Patel, R. P. (1969). An experimental study of the thermal wake interference between closely spaced wires of an x-type hot wire probe. McGill University, Mechanical Engineering Research Laboratory Report 69-7.

Gunkel, A. A., Patel, R. P., and Weber, M. E. (1971). A shielded hot-wire probe for highly turbulent flows and rapidly reversing flows. *Ind. Engr. Chem. Fund.*, 10, 627–631.

Gupta, A. K., and Srivastava, A. (1979). Feasibility study of a reverse flow sensing probe. *J. Phys. E.: Sci. Inst.*, 12, 1029–1030.

Guy, T. B. (1971). Gas velocity measurements in a shock tube with a hot wire anemometer. *J. Phys. E.: Sci. Inst.*, 4, 961–965.

Hah, C., and Lakshminarayana, B. (1978). Effect of rotation on a rotating hot-wire sensor. *J. Phys. E.: Sci. Inst.*, 11, 999–1001.

Hartmann, U. (1982). Wall interference effects on hot-wire probes in a nominally two-dimensional highly curved wall jet. *J. Phys. E.: Sci. Inst.*, 15, 725–730.

Hatton, A. P., James, D. D., and Swire, H. W. (1970). Combined forced and natural convection with low speed air flow over horizontal cylinders. *J. Fluid Mech.*, 42, 17–31.

Hauptmann, E. G. (1968). A simple hot wire anemometer probe. *J. Phys. E.: Sci. Inst.*, 1, 874–875.

Hembling, D. (1980). Tune in the wind – a do-it-yourself hot-wire anemometer. *73 Mag.*, November, 80–84.

Henry, R. M., and Greene, G. C. (1974). Anemometers for Mars. In *Flow, Its Measurement and Control in Science and Industry* (R. B. Dowdell, Ed.), Vol. 1, pp. 633–638. Instrument Society of America, Pittsburgh.

Herringe, R. A., and Davis, M. R. (1974). Detection of instantaneous phase changes in gas–liquid mixtures. *J. Phys. E.: Sci. Inst.*, 7, 807–812.

Herzog, S., and Lumley, J. L. (1978). Determination of large eddy structures in the viscous sublayer: A progress report. In *Proc. Dyn. Flow Conf.*, pp. 869–885. Skovlunde, Denmark.

Hill, J. C., and Sleicher, C. A. (1971). Directional sensitivity of hot film sensors in liquid metals. *Rev. Sci. Instr.*, 42, 1461–1468.

Hinze, J. O. (1959). *Turbulence*. McGraw-Hill, New York.

Hoff, M. (1968). Hot-film anemometry techniques in liquid mercury. Grumman Res. Dept., Memo RM-414J.

Hoffmeister, M. (1972). Using a single hot-wire probe in three-dimensional turbulent flow fields. *DISA Info.*, 13, 26–28.

Hojstrup, J., Rasmussen, K., and Larsen, S. E. (1976). Dynamic calibration of temperature wires in still air. *DISA Info.*, 20, 22–30.

   (1977). Dynamic calibration of temperature wires in moving air. *DISA Info.*, 21, 33.

Holroyd, R. J. (1979). An experimental study of the effects of wall conductivity, non-uniform magnetic fields and variable-area ducts on liquid metal flows at high Hartmann number. Part 1. Ducts with non-conducting walls. *J. Fluid Mech.*, 93, 609–630.

Hussain, A. K. M. F., and Zaman, K. B. M. Q. (1978). The free shear layer tone phenomena and probe interference. *J. Fluid Mech.*, 87, 349–383.

Hutton, P. and Gammon, L. N. (1976). The effect of probe-body temperature changes on the behavior of flush-mounting film probes. *DISA Info.*, 20, 4.

Jacobsen, R. A. (1977). Hot-wire anemometry for in-flight measurement of aircraft wake vortices. *DISA Info.*, 21, 21–27.

James, D. F., and Acosta, A. J. (1970). The laminar flow of dilute polymer solutions around circular cylinders. *J. Fluid Mech.*, 42, 269–288.

Jerome, F. E., Guitton, D. E., and Patel, R. P. (1971). Experimental studies of the thermal wake interference between closely spaced wires of a x-type hot-wire probe. *Aero. Quart.*, 22, 119–126.

Jiminez, J., Martinez-Val, R., and Rebollo, M. (1981). Hot-film sensors calibration and drift in water. *J. Phys. E.: Sci. Inst.*, 14, 569–572.

Johnston, J. B., and Llewellyn, F. B. (1934). Limits to amplification. *Elect. Engr.*, 53, 1449–1454.

Jorgensen, F. E. (1971). Directional sensitivity of wire and fiber-film probes. *DISA Info.*, 11, 31–37.

   (1979). An omnidirectional thin-film probe for indoor climate research. *DISA Info.*, 24, 24–29.

   (1982). Characteristics and calibration of a triple-split probe for reversing flows. *DISA Info.*, 27, 15–22.

Kalashnikov, V. N., and Kudin, A. M. (1973). Calibration of hot-film probes in water and in polymer solutions. *DISA Info.*, 14, 15–18.

Kanevce, G., and Oka, S. (1973). Correcting hot-wire readings for influence of fluid temperature variations. *DISA Info.*, 15, 21–24.

Kastrinakis, E. G., Eckelmann, H., and Willmarth, W. W. (1979). Influence of the flow velocity on a Kovasznay type vorticity probe. *Rev. Sci. Instr.*, 50, 759–767.

Kastrinakis, E. G., Wallace, J. M., Willmarth, W. W., Ghorashi, B., and Brodkey, R. S. (1977). On the mechanism of bounded turbulent shear flows. In *Lecture Notes in Physics*, Vol. 75, pp. 175–189.

Keesee, J. E., Francis, M. S., and Lang, J. D. (1979). Technique for vorticity measurement in unsteady flow. *AIAA J.*, 17, 387–393.

Kidron, I. (1966). Application of modulated electromagnetic waves for measurement of the frequency response of heat-transfer transducers. *DISA Info.*, 4, 25–29.

Kielbasa, J., Rysz, J., Smolarski, A. Z., and Stasicki, B. (1972). The oscillatory anemometer. In *Fluid Dynamic Measurements* (D. J. Cockrell, Ed.) pp. 65–67.

King, L. V. (1914). On the convection of heat from small cylinders in a stream of fluid: Determination of the convection constants of small platinum wires with applications to hot-wire anemometry. *Phil. Trans. Roy. Soc. A*, 214, 373–432.

Kirchoff, R. H., and Struziak, R. M. (1976). Direct measurement of the mean flow velocity vector. *J. Fluids Engr.*, 98, 736–739.

Kistler, A. L. (1959). Fluctuation measurements in a supersonic turbulent boundary layer. *Phys. Fluids*, 2, 290–296.

Ko, C. L., McLaughlin, D. K., and Troutt, T. R. (1978). Supersonic hot-wire fluctuation data analysis with a conduction end-loss correction. *J. Phys. E.: Sci. Inst.*, 11, 488–494.

Ko, N. W. M., and Davies, P. O. A. L. (1971). Interference effect of hot wires. *IEEE Trans. Instr. Meas.*, 20, 76–78.

Koch, F. A., and Gartshore, I. S. (1972). Temperature effects on hot wire anemometer calibrations. *J. Phys. E.: Sci. Inst.*, 5, 58–61.

Kovasznay, L. S. G. (1949). Hot wire investigation of the wake behind cylinders at low Reynolds numbers. *Proc. Roy. Soc. A*, 198, 174–190.

(1950). The hot-wire anemometer in supersonic flows. *J. Aero. Sci.*, 17, 565–573.

(1953). Turbulence in supersonic flow. *J. Aero. Sci.*, 20, 657–674.

(1954). Hot wire method. In *Physical Measurements in Gasdynamics and Combustion* (R. Ladenburg, Ed.), Vol. 9, pp. 219–276. Princton University Press, Princeton, New Jersey.

Kramers, H. (1946). Heat transfer from spheres to flowing media. *Physica*, 12, 61–120.

Laderman, A. J., and Demetriades, A. (1973). Hot-wire measurements of hypersonic boundary-layer turbulence. *Phys. Fluids*, 16, 179–181.

(1974). Mean and fluctuating flow measurements in the hypersonic boundary layer over a cooled wall. *J. Fluid Mech.*, 63, 121–144.

Larsen, S. E., and Busch, N. E. (1974). Hot-wire measurements in the atmosphere. Part 1: Calibration and response characteristics. *DISA Info.*, 16, 15–33.

(1976). Hot-wire measurements in the atmosphere. Part 2: A field experiment in the surface boundary layer. *DISA Info.*, 20, 5–21.

Laufer, J., and McClellan, R. (1956). Measurement of heat transfer from fine wires in supersonic flows. *J. Fluid Mech.*, 1, 276–289.

Libby, P. A. (1977). Studies in variable-density and reacting turbulent shear flows. In *Studies in Convection* (B. E. Launder, Ed.), Vol. 2, pp. 1–43. Academic Press, New York.

Liepmann, H., and Skinner, G. (1954). Shearing-stress measurements by use of a heated element. NACA TM 3268.

Ling, F. F., and Lowe, G. (1981). A technique for low velocities measurements. *J. Phys. E.: Sci. Inst.*, 14, 932–933.

Ling, S. C. (1955). Measurement of flow characteristics by the hot film technique. Ph.D. thesis, University of Iowa.

Ling, S. C., Atabeck, H. B., Fry, D. L., Patel, D. J., and Janicki, J. S. (1968). Application of heated-film velocity and shear probes to hemodynamic studies. *Circ. Res.*, 23, 789–801.

Lowell, H. H. (1950). Design and applications of hot wire anemometers for steady state measurements at transonic and supersonic airspeeds. NACA TN 2117.

Lu, S. S. (1979). Dynamic characteristics of a simple constant-temperature hot-wire anemometer. *Rev. Sci. Instr.*, 50, 772–775.

Ludwieg, H. (1950). Instrument for measuring the wall shearing stress of turbulent boundary layers. NACA TM 1284.

Lueck, R. G., and Osborn, T. R. (1979). On the frequency response of wire sensors as measured by internal and acoustic heating. *DISA Info.*, 24, 36–40.

McAdams, W. H. (1954). *Heat Transmission*, 3d ed. McGraw-Hill, New York.

McConachie, P. J., and Bullock, K. J. (1975). Simultaneous measurement of longitudinal velocity and temperature fluctuations with a single wire in non-isothermal flow. In *Turbulence in Liquids* (G. K. Patterson and J. L. Zakin, Eds.), pp. 14–25. University of Missouri, Rolla.

McQuaid, J., and Wright, W. (1973). The response of a hot-wire anemometer in flows of gas mixtures. *Int. J. Heat Mass Trans.*, 16, 819–828.

McQuivey, R. S. (1972). Principles and measuring techniques of turbulence characteristics in open-channel flows. U. S. Dept. Interior, Geological Survey, Bay St. Louis, Mississippi, Paper 802-A, July.

Mahler, D. S. (1982). Bidirectional hot-wire anemometer. *Rev. Sci. Instr.*, 53, 1465–1466.

Malcolm, D. G. (1969). Some aspects of turbulence measurement in liquid mercury using cylindrical quartz-insulated hot-film sensors. *J. Fluid Mech.*, 37, 701–713.

    (1970). An investigation of the stability of a magnetohydrodynamic shear layer. *J. Fluid Mech.*, 41, 531–544.

Mestayer, P., and Chambaud, P. (1979). Some limitations to measurements of turbulence micro-structure with hot and cold wires. *Bound. Layer Met.*, 16, 311–329.

Metzner, A. B., and Astarita, G. (1967). External flow of viscoelastic materials: Fluid property restrictions on the use of velocity-sensitive probes. *AIChE J.*, 13, 550–555.

Mikulla, V., and Horstman, C. C. (1975). Turbulence stress measurements in a nonadiabatic hypersonic boundary layer. *AIAA J.*, 13, 1607–1613.

Miller, C. H. (1972). A three degree of freedom traversing gear. *DISA Info.*, 13, 34–36.

Miller, G. E. (1980). Position sensitivity of hot-film shear probes. *J. Phys. E.: Sci. Inst.*, 13, 973–976.

Ming Ho, C. (1982). Response of a split film probe under electrical perturbations. *Rev. Sci. Instr.*, 53, 1240–1245.

Ming Ho, C., Jakus, K., and Parker, K. H. (1976). Temperature fluctuations in a turbulent flame. *Comb. Flame*, 27, 113–123.

Moffat, R. J., Yavuzkurt, S., and Crawford, M. E. (1978). Real-time measurements of turbulence quantities with a triple hot-wire system. In *Proc. Dyn. Flow Conf.*, pp. 1013–1035. Skovlunde, Denmark.

Mojola, O. O. (1974). A hot-wire method for three-dimensional shear flows. *DISA Info.*, 16, 11–14.

Mollenkopf, G. (1972). Measuring nonstationary periodical flow in the draft tube of a water-powered Francis model turbine. *DISA Info.*, 13, 11–15.

Morkovin, M. V. (1956). Fluctuations and hot wire anemometry in compressible flows. AGARDograph 24.

Mulhearn, P. J., and Finnigan, J. J. (1978). A simple device for dynamic testing of x-configuration hot-wire anemometer probes. *J. Phys. E.: Sci. Inst.*, 11, 679–681.

Nerem, R. M., Seed, W. A., and Wood, N. B. (1972). An experimental study of the velocity distribution and transition to turbulence in the aorta. *J. Fluid Mech.*, 52, 137–160.

Neuerburg, W. (1969). Directional hot-wire probe. *DISA Info.*, 7, 30–31.

Oka, S. and Kostic, Z. (1972). Influence of wall proximity on hot-wire velocity measurements. *DISA Info.*, 13, 29–33.

Owen, F. K., and Bellhouse, B. J. (1970). Skin-friction measurement of supersonic speed. *AIAA J.*, 8, 1358–1360.

Parthasarathy, S. P., and Tritton, D. J. (1963). Impossibility of linearizing a hot-wire anemometer for measurements in turbulent flows. *AIAA J.*, 1, 1210–1211.

Patel, R. P. (1963). Measurements of Reynolds stresses in a circular pipe as a means of testing a DISA constant-temperature hot wire-anemometer. McGill University Tech. Note 63-6.

Patterson, A. M. (1958). Turbulence spectrum studies in the sea with hot wires. Pacific Naval Lab., Esquimalt, B.C., Canada.

Patty, F. A., Ed. (1981). *Patty's Industrial Hygiene and Toxicology*, 3d rev. ed. Vol. 2A, pp. 1775–1789. Interscience, New York.

Perry, A. E. (1977). The time response of an aspirating probe in gas sampling. *J. Phys. E.: Sci. Inst.*, 10, 898–902.

Perry, A. E., and Morrison, G. L. (1971). A study of the constant-temperature hot-wire anemometer. *J. Fluid Mech.*, 47, 577–599.

(1971a). Static and dynamic calibrations of constant-temperature hot-wire systems. *J. Fluid Mech.*, 47, 765–777.

(1971b). Vibration of hot-wire anemometer filaments. *J. Fluid Mech.*, 50, 815–825.

Pichon, J. (1970). Comparison of some methods of calibrating hot-film probes in water. *DISA Info.*, 10, 15–21.

Platnieks, I. A. (1971). Comparison of the hot-wire anemometric and conduction methods of measuring velocity characteristics of a flow of mercury in a transverse magnetic field. *Mag. Gidro.*, 7, 140–142.

Polyakov, A. F., and Shindin, S. A. (1978). Peculiarities of hot-wire measurements of mean velocity and temperature in the wall vicinity. *Lett. Heat Mass Trans.*, 5, 53–58.

Rajasooria, G. P. D., and Brundin, C. L. (1971). Use of hot wires in low density flows. *AIAA J.*, 9, 979–981.

Rasmussen, C. G. (1967). The air bubble problem in water flow hot film anemometry. *DISA Info.*, 5, 21–26.

Rey, C., and Beguier, C. (1977). On the use of a three parallel wire probe. *DISA Info.*, 21, 11–15.

Richardson, E. V., and McQuivey, R. S. (1968). Measurement of turbulence in water. *J. Hyd.*, 94, 411–430.

Robinson, T., and Larsson, K. (1973). An experimental investigation of a magnetically driven rotating liquid metal flow. *J. Fluid Mech.*, 60, 641–664.

Roos, F. W., and Bogar, T. J. (1982). Comparison of hot-film probe and optical techniques for sensing shock motion. *AIAA J.*, 20, 1071–1076.

Roshko, A. (1954). On the development of turbulent wakes from vortex streets. NACA Report 1191.

Rubatto, G. (1970). Calibration of probes for flow velocity measurements in liquids, in the range 2–5 m/sec. *DISA Info.*, 9, 3–7.

Sajben, M. (1965). Hot wire anemometer in liquid mercury. *Rev. Sci. Instr.*, 36, 945–949.

Sakao, F. (1973). Constant temperature hot wires for determining velocity fluctuations in an air flow accompanied by temperature fluctuations. *J. Phys. E.: Sci. Inst.*, 6, 913–916.

Sandborn, V. A. (1972). *Resistance Temperature Transducers*. Metrology Press, Ft. Collins, Colorado.

Sawatzki, O. (1971). Measurements in the disturbed laminar boundary layer of a special three-dimensional flow field. *DISA Info.*, 11, 5–24.

Schacher, G., and Fairall, C. W. (1976). Use of resistance wires for atmospheric turbulence measurements in the marine environment. *Rev. Sci. Instr.*, 47, 703–707.

    (1979). Frequency response of cold wires used for atmospheric turbulence measurements in the marine environment. *Rev. Sci. Instr.*, 50, 1463–1466.

Schmidt, E. M., and Cresci, R. J. (1971). Hot-wire anemometry in low-density flows. *AIAA J.*, 9, 2019–2027.

Schubauer, G. B. (1935). Effect of humidity in hot-wire anemometry. *J. Res. Nat. Bureau Standards*, 15, 575–578.

Schubauer, G. B., and Klebanoff, P. S. (1946). Theory and application of hot wire instruments in the investigation of turbulent boundary layers. NACA ACR 5K27 (W-86).

Seed, W. A. (1969). Fabrication of thin-film microcircuits on curved substrates. *J. Phys. E.: Sci. Inst.*, 2, 206.

Seed, W. A., and Thomas, I. R. (1972). The application of hot-film anemometry to the measurement of blood flow velocity in man. In *Fluid Dynamic Measurements in the Industrial and Medical Environments* (D. J. Cockrell, Ed.), pp. 298–304. Leicester University Press.

Seed, W. A., and Wood, N. B. (1969). An apparatus for calibrating velocity probes in liquids. *J. Phys. E.: Sci. Inst.*, 2, 896–898.

    (1970). Development and evaluation of a hot-film velocity probe for cardiovascular studies. *Cardio. Res.*, 4, 253–263.

    (1970a). Use of a hot-film velocity probe for cardiovascular studies. *J. Phys. E.: Sci. Inst.*, 3, 377–384.

Shapiro, A. H. (1953). *The Dynamics and Thermodynamics of Compressible Fluid Flow*. Ronald Press, New York.

Shemdin, O. (1969). Instantaneous velocity and pressure measurements above propagating waves. University of Florida, Dept. Coastal and Oceanographic Engineering Report 4.

Shepard, C. E. (1955). A self-excited alternating-current constant-temperature hot-wire anemometer. NACA TN 3406.

Shiralkar, B. S. (1970). Local void fraction measurements in Freon-114 with a hot-wire anemometer. General Electric Co., NEDO-13158.

Siddall, R. G., and Davies, T. W. (1972). An improved response equation for hot-wire anemometry. *Int. J. Heat Mass Trans.*, 15, 367–368.

Smith, K. A., Merrill, E. W., Mickley, H. S., and Virk, P. S. (1967). Anomalous pitot tube and hot film measurements in dilute polymer solutions. *Chem. Engr. Sci.*, 22, 619–626.

Smits, A. J., Perry, A. E., and Hoffmann, P. H. (1978). The response to temperature fluctuations of a constant-current hot-wire anemometer. *J. Phys. E.: Sci. Inst.*, 11, 909–914.

Spangenberg, W. G. (1955). Heat loss characteristics of hot wire anemometers at various densities in transonic and supersonic flow. NACA TN 3381.

Sreenivasan, K., and Ramachandran, A. (1961). Effect of vibration on heat transfer from a horizontal cylinder to a normal air stream. *Int. J. Heat Mass Trans.*, 3, 60–67.

Stevens, R. G., Borden, A., and Strausser, P. E. (1956). Summary report on the development of a hot wire turbulence sensing element for use in water. David Taylor Model Basin Report 953.

Strohl, A., and Comte-Bellot, G. (1973). Aerodynamic effects due to configuration of x-wire anemometers. *J. Appl. Mech.*, 40, 661–666.

Tillmann, W., and Haubinger, G. (1978). Wall shear stress measurements in artificial heart valves with hot-film probes. In *Proc. Dyn. Flow Conf.*, pp. 737–743. Skovlunde, Denmark.

Tombach, I. H. (1973). An evaluation of the heat pulse anemometer for velocity measurement in inhomogeneous turbulent flow. *Rev. Sci. Instr.*, 44, 141–148.

Toral, H. (1981). A study of the hot-wire anemometer for measuring void fraction in two phase flow. *J. Phys. E.: Sci. Inst.*, 14, 822–827.

Townsend, A. A. (1951). The diffusion of heat spots in isotropic turbulence. *Proc. Roy. Soc. A*, 209, 418–430.

Tritton, D. J. (1967). Note on the effect of a nearby obstacle on turbulence intensity in a boundary layer. *J. Fluid Mech.*, 28, 433–437.

TSI, Inc. catalog, 1978.

Van der Hegge Zijnen, B. G. (1956). Modified correlation formulae for the heat transfer by natural and by forced convection from horizontal cylinders. *Appl. Sci. Res. A*, 6, 129–140.

van Thinh, N. (1969). On some measurements made by means of a hot wire in a turbulent flow near a wall. *DISA Info.*, 7, 13–18.

Vidal, R. J., and Golian, T. C. (1967). Heat-transfer measurements with a catalytic flat plate in dissociated oxygen. *AIAA J.*, 5, 1579–1588.

Vonnegut, B., and Neubauer, R. (1952). Detection and measurement of aerosol particles. *Anal. Chem.*, 24, 1000–1005.

Vukoslavcevic, P., and Wallace, J. M. (1981). Influence of velocity gradients on measurements of velocity and streamwise vorticity with hot-wire x-array probes. *Rev. Sci. Instr.*, 52, 869–879.

Walker, R. E., and Westenberg, A. A. (1956). Absolute low speed anemometer. *Rev. Sci. Instr.*, 27, 844–848.

Walker, T. B., and Bullock, K. J. (1972). Measurement of longitudinal and normal velocity fluctuations by sensing the temperature downstream of a hot wire. *J. Phys. E.: Sci. Inst.*, 5, 1173–1178.

Wasan, D. T., and Baid, K. M. (1971). Measurement of velocity in gas mixtures: Hot-wire and hot-film anemometry. *AIChE J.*, 17, 729–731.

Way, J., and Libby, P. A. (1970). Hot-wire probes for measuring velocity and concentration in helium–air mixtures. *AIAA J.*, 8, 976–978.

Webster, C. A. G. (1962). A note on the sensitivity to yaw of a hot-wire anemometer. *J. Fluid Mech.*, 13, 307–312.

Wehrmann, O. H. (1968). Recent developments in hot-wire techniques. In *Advances in Hot Wire Anemometry* (W. L. Melnik and J. R. Weske, Eds.), pp. 194–202. University of Maryland, College Park.

Weidman, P. D., and Browand, F. K. (1975). Analysis of a simple circuit for constant temperature anemometry. *J. Phys. E.: Sci. Inst.*, 8, 553–560.

Willmarth, W. W. (1978). Nonsteady vorticity measurements: Survey and new results. In *Proc. Dyn. Flow Conf.*, pp. 1003–1012. Skovlunde, Denmark.

Willmarth, W. W., and Bogar, T. J. (1977). Survey and new measurements of turbulent structure near a wall. *Phys. Fluids*, 20, S15–S16.

Wills, J. A. B. (1962). The correction of hot-wire readings for proximity to a solid boundary. *J. Fluid Mech.*, 12, 388–396.

    (1976). A submerging hot wire for flow measurement over waves. *DISA Info.*, 20, 31–34.

Wooldridge, C. E., and Muzzy, R. J. (1966). Boundary-layer turbulence measurements with mass addition and combustion. *AIAA J.*, 4, 2009–2016.

Wygnanski, I., and Ho, C. (1978). Note on the prong configuration of an x-array hot wire probe. *Rev. Sci. Instr.*, 49, 865–866.

Wyngaard, J. C., and Lumley, J. L. (1967). A constant temperature hot-wire anemometer. *J. Sci. Instr.*, 44, 363–365.

# INDEX

209